In Memorium
Serge Lang
1927–2005

MEMOIRS
of the
American Mathematical Society

Number 946

Heat Eisenstein Series on $SL_n(C)$

Jay Jorgenson
Serge Lang

September 2009 • Volume 201 • Number 946 (end of volume) • ISSN 0065-9266

American Mathematical Society
Providence, Rhode Island

2000 *Mathematics Subject Classification.*
Primary 35K05, 58J35, 11F72; Secondary 11M36.

Library of Congress Cataloging-in-Publication Data

Jorgenson, Jay.
Heat Eisenstein series on $\mathrm{SL}_n(C)$ / Jay Jorgenson and Serge Lang.
p. cm. — (Memoirs of the American Mathematical Society, ISSN 0065-9266 ; no. 946)
"Volume 201, number 946 (fifth of 5 numbers)."
Includes bibliographical references and index.
ISBN 978-0-8218-4044-3 (alk. paper)
1. Heat equation. 2. Eisenstein series. 3. Decomposition (Mathematics) 4. Function spaces. I. Lang, Serge. II. Title.

QA377.J657 2009
515′.353—dc22 2009019838

Memoirs of the American Mathematical Society

This journal is devoted entirely to research in pure and applied mathematics.

Subscription information. The 2009 subscription begins with volume 197 and consists of six mailings, each containing one or more numbers. Subscription prices for 2009 are US$709 list, US$567 institutional member. A late charge of 10% of the subscription price will be imposed on orders received from nonmembers after January 1 of the subscription year. Subscribers outside the United States and India must pay a postage surcharge of US$65; subscribers in India must pay a postage surcharge of US$95. Expedited delivery to destinations in North America US$57; elsewhere US$160. Each number may be ordered separately; *please specify number* when ordering an individual number. For prices and titles of recently released numbers, see the New Publications sections of the *Notices of the American Mathematical Society.*

Back number information. For back issues see the *AMS Catalog of Publications.*

Subscriptions and orders should be addressed to the American Mathematical Society, P. O. Box 845904, Boston, MA 02284-5904 USA. *All orders must be accompanied by payment.* Other correspondence should be addressed to 201 Charles Street, Providence, RI 02904-2294 USA.

Memoirs of the American Mathematical Society (ISSN 0065-9266) is published bimonthly (each volume consisting usually of more than one number) by the American Mathematical Society at 201 Charles Street, Providence, RI 02904-2294 USA. Periodicals postage paid at Providence, RI. Postmaster: Send address changes to Memoirs, American Mathematical Society, 201 Charles Street, Providence, RI 02904-2294 USA.

This publication is indexed in *Science Citation Index*®, *SciSearch*®, *Research Alert*®,
CompuMath Citation Index®, *Current Contents*®/*Physical, Chemical & Earth Sciences.*
Printed in the United States of America.

∞ The paper used in this book is acid-free and falls within the guidelines
established to ensure permanence and durability.
Visit the AMS home page at `http://www.ams.org/`

10 9 8 7 6 5 4 3 2 1 14 13 12 11 10 09

Contents

ABSTRACT. The purpose of this Memoir is to define and study multi-variable Eisenstein series attached to heat kernels. Fundamental properties of heat Eisenstein series are proved, and conjectural behavior, including their role in spectral expansions, are stated.

Recieved by the editor 5-26-2002 and in revised form 4-26-2004.

2000 *Mathematics Subject Classification.* Primary 35K05, 58J35, 11F72; Secondary 11M36.

The first author was supported in part by numerous NSF Grant and PSC-CUNY awards.

Key words and phrases. Heat kernels, Eisenstein series, spectral expansions.

Acknowledgements

Jorgenson thanks PSC-CUNY and the NSF for grant support. Lang thanks Tony Petrello for his support of the Yale Math Department and of our joint work. Lang also thanks the Max Planck Institute for productive yearly visits. The main conjectures in this monograph were presented by Lang on 12 November 1997 in a seminar lecture at the Max Planck Institute. We thank Mel DelVecchio for her patience in setting the manuscript in TEX, in a victory of person over machine, and Andy Sinton for his careful reading and exhaustive list of comments.

Jay Jorgenson and Serge Lang
April 2004

The final proofreading and preparation of this monograph took place after the passing of Serge Lang. As I have stated elsewhere, since the 1990's Serge Lang became fascinated with the prospect of using heat kernels and heat kernel analysis in number theory. Lang and I worked together on mathematics for nearly 15 years. During the course of our many, many conversations, we developed a long-term program of study which would create general zeta functions using heat kernels. The present monograph contains some of the goals, ideas, and dreams which we envisioned. In additional to mathematics itself, Lang was well-known for his style of exposition. As a result, I have chosen to not alter the manuscript after Lang's passing, except for typesetting considerations.

Jay Jorgenson
May 2009

Introduction

The general context for this work is the spectral decomposition (Fourier decomposition) of functions on spaces of the form $\Gamma\backslash G/K$, where G is a semisimple Lie group (or reductive group), K is a maximal compact subgroup and Γ is an arithmetic discrete subgroup. The general theory has few references, and all of them are difficult to get into, for various reasons. Aside from the pioneering work of Selberg, the theory made a great advance in Langlands [**Lgds 66**], [**Lgds 76**]. However, Langlands himself never rewrote what was only a draft of a huge project. He himself stated in [**Lgds 76**]: "I myself now have difficulty finding my way through it", referring to section 7, which is at the heart of the matter. (See page 284.)

Since $\Gamma\backslash G/K$ is not compact, the main difficulties lie in the continuous spectrum, which is determined by the Eisenstein series. These are defined by sums over subsets of Γ, and these sums converge in a certain domain, just as classical Dirichlet series converge in a half plane. One problem is to extend them meromorphically to a complex space, usually of dimension > 1, and then to prove a functional equation. In 1968, Harish-Chandra carried out a more complete exposition of the Langlands theory (still unpublished at the time), but his treatment although very useful, was short of a stable definitive version for several reasons. First, he limited himself to the spectral decomposition of what are called cusp forms. The space $L^2(\Gamma\backslash G/K)$ splits into an orthogonal sum of the cuspidal part and what is called the residual part. Langlands shows how the latter can be derived (non-trivially) from the former by the method of "iterated residues" in the "jungle" of that section 7, as Langlands himself says. It's rough going.

Secondly, [**Har 68**] is written for experts who are well acquainted with an extensive background in Lie groups and Lie algebras, and a number of background results are assumed, or left to the reader as having "easy" proofs, without any reference to what to experts is standard basic material.

An attempt by Moeglin-Waldspurger [**MoW 94**] to expand and complete Langlands' 1976 Springer Lecture Notes is valuable for some, but for someone like us coming into the subject from the outside, the evaluation by Laumon, who was one of the participants of the seminar from which [**MoW 94**] evolved, applies: "... the goal of the seminar was to render obscure what was already not so clear." (See the first page of the introduction to [**MoW 94**].)

One expository problem which arises over 30 years has been the insistence of the Lie industry to have expositions carrying out the general case ("Let G be a semisimple Lie group..." or "Let G be a reductive group..."). A number of the difficulties which Langlands lists in Appendix III of [**Lgds 76**] have to do precisely with the need to treat general such groups. Cf. [**Lgds 76**], p. 284.

Another problem, as in [**MoW 94**] lies in the adelization, which mixes two number theoretic aspects, and to some extent obscures certain aspects of analysis. Fortunately for us, Harish-Chandra in [**Har 68**] was not yet adelized. On the other hand, he carries out only the "cuspidal" case, and there is no similar exposition for the "residual" case.

Major progress beyond Langlands and Harish-Chandra took place in papers of Werner Müller, who carried out the spectral decomposition in the residual case [**Mue 83**], [**Mue 00**], and proved some of the outstanding conjectures [**Mue 89**], [**Mue 98**]. However, Müller refers to Langlands and Harish-Chandra as the need arises, so inexperienced readers like us still have a hard time getting into his papers.

We come to the theory as outsiders. Roughly speaking, we are getting at zeta functions via the following route [**JoL 93**], [**JoL 94**], [**JoL 01c**].

- We start with the heat kernel, say on G/K.
- We periodize it, getting the heat kernel $\mathbf{K}^{\Gamma}(t,x,y)$ invariant under Γ in each variable $x, y \in G/K$ (never mind the singularities for the moment).
- We derive a Fourier decomposition of the heat kernel on $\Gamma\backslash G/K$ which amounts to a theta inversion relation.
- We apply the Gauss transform [**JoL 94**] to a regularized trace to get what amounts to a functional equation of what amounts to a zeta function.

Actually, what one gets by this procedure is the logarithmic derivative of something which deserves to be called a zeta function. In the simplest case of a compact Riemann surface, it would turn out to be the Selberg zeta function according to a theorem of McKean.

To carry out in detail the above program would (will) take several papers, Lecture Notes, and books. What we want to see is one typical special case which minimizes all difficulties having to do with algebraic geometry, differential geometry, and Lie theory. Thus we choose the special case of SL_n. It is here essential to deal with arbitrary n. The case of SL_2 is too special, and does not have enough structure by itself to reflect all the properties we want to reflect. Conjecturally, it occurs merely at the bottom of the SL_n-ladder, and the way zeta functions at lower levels occur as fudge factors in zeta functions of higher levels is one of the main features which affects all levels, including the bottom level of SL_2, or even one level lower which is the level of the classical zeta functions of analytic number theory, including the Riemann zeta function. For more detailed comments on this situation, see the last "Onward" section of Chapter 4 which can be viewed as a continuation of this introduction. However, these additional comments are best placed after the statement of the conjectural theta relations in Chapter 4, Section 6, formula **SPEX 1**.

Even with SL_n, we have a choice whether to consider $\mathrm{SL}_n(\mathbf{R})$ or $\mathrm{SL}_n(\mathbf{C})$. A priori, $\mathrm{SL}_n(\mathbf{R})$ is slightly simpler, and if one considers only G/K, that is if one considers only the theory of spherical functions without Γ, then $\mathrm{SL}_n(\mathbf{R})$ provides an appropriate introductory context for spherical inversion following Harish-Chandra, later simplified by Helgason, Gangolli, Rosenberg and Anker. We gave an exposition in [**JoL 01a**], which serves as background for the next step involving the discrete group Γ. For this step, it turns out that $\mathrm{SL}_n(\mathbf{C})$ is simpler because the heat kernel in the complex case is "split", i.e. has a structure whereby it looks just like the heat kernel on euclidean space, essentially the gaussian function suitably normalized, which on G/K is the inverse spherical transform of the standard gaussian function on euclidean space. Thus we resolutely write the present work for $\mathrm{SL}_n(\mathbf{C})$, hoping to make the whole theory more accessible to people who are not experts in Lie theory. Even then, plenty of serious difficulties remain to be faced. Thus the present work is just the first four chapters of what we expect to be open ended.

Notation and Terminology

We follow the conventions of [**JoL 01a**], where readers will find the background material and proofs.

For concreteness we let $G = \mathrm{SL}_n(\mathbf{C})$, and:

U = upper triangular unipotent matrices
A = diagonal matrices with positive diagonal elements, determinant 1.
K = maximal complex unitary subgroup.

We should really have an index G, and so write U_G, A_G, K_G, but as long as G is fixed we omit the index.

We have the **Iwasawa decomposition**

$$G = UAK,$$

and the product map $U \times A \times K \to UAK$ is a differential isomorphism. Furthermore, let $\mathrm{Pos}_n = \mathrm{Pos}_n(\mathbf{C})$ = space of positive definite hermitian matrices. Then the map

$$x \mapsto xx^* \qquad \text{with} \quad x^* = {}^t\bar{x}$$

is a differential isomorphism of $G/K \xrightarrow{\approx} \mathrm{SPos}_n$, so of $UA \xrightarrow{\approx} \mathrm{SPos}_n$.

We recall the proof briefly. Let $\{e_1, \dots e_n\}$ be the standard unit vectors of $\mathbf{C}^n$. Let $x \in G$. Let $v_i = xe_i$. We orthogonalize $\{v_1, \dots, v_n\}$ by the standard Gram-Schmidt process, so we use a transformation by a matrix $u \in U$, namely, we let

$$w_1 = v_1, \qquad w_2 = v_2 - c_{21}w_1 \perp w_1,$$
$$w_3 = v_3 - c_{32}w_2 - c_{31}w_1 \perp w_1 \text{ and } w_2,$$
and so on.

Then $e'_i = w_i/\|w_i\|$ (standard hermitian norm on $\mathbf{C}^n$) is a unit vector, and the matrix a having $a_i = \|w_i\|^{-1}$ for its diagonal elements is in A. Let

$$k = aux \quad \text{so} \quad x = u^{-1}a^{-1}k.$$

Then k is unitary, which proves $G = UAK$. To show uniqueness, we use the map $x \mapsto xx^*$. Note that $xx^* = I$ (identity) if and only if $x \in K$ (x is unitary). Suppose that

$$u_1a^2u_1^* = u_2b^2u_2^* \qquad \text{with } u_1, u_2 \in U \text{ and } a, b \in A.$$

Putting $u = u_2^{-1}u_1$ we find $ua = bu^*$. Since u and u^* are triangular in opposite directions, they must be diagonal and finally $a^2 = b^2$ so $a = b$ because the diagonal components are positive. This proves uniqueness. The differential isomorphism property is proved by the computation of a Jacobian, which we do not reproduce here. Cf. [**JoL 01a**], Chapter I, §2.

Given $x = uak \in G$, with its Iwasawa decomposition, we let

$$x_A = a$$

denote its A-projection, and similarly for x_U and x_K. We also denote these respectively by

$$\mathrm{Iw}_A(x), \qquad \mathrm{Iw}_U(x), \qquad \mathrm{Iw}_K(x).$$

Let $\mathfrak{n} = \mathrm{Lie}(U)$ be the Lie algebra of U, as complex vector space, so $\mathfrak{n}$ is also a real vector space of twice the complex dimension. Let $\mathfrak{a} = \mathrm{Lie}(A), \mathfrak{g} = \mathrm{Lie}(G)$, so $\mathfrak{g}$ acts on itself by the **Lie regular representation**, that is for $X \in \mathfrak{g}$, $Z \in \mathfrak{g}, X_{\mathfrak{g}}$ or $[X]$ is the linear endomorphism of $\mathfrak{g}$ such that

$$X_{\mathfrak{g}}(Z) = [X, Z] = XZ - ZX.$$

An element $x \in G$ acts on G by conjugation, $\mathbf{c}(x)y = xyx^{-1}$. It then acts functorially on any functor. In particular, it acts on the Lie algebra $\mathfrak{g}$, by what we also call the **conjugation representation**, that is

$$\mathbf{c}_{\mathrm{Lie}}(x)Z = xZx^{-1} = Z^{x^{-1}}.$$

We often omit the subscript and write simply $\mathbf{c}(x)Z = xZx^{-1}$.

The $(\mathfrak{a}, \mathfrak{n})$-representation: semisimplicity. Under the Lie regular representation of $\mathfrak{a}$ on $\mathfrak{g}$ (and also under the conjugation representation restricted to A), the Lie algebra $\mathfrak{g}$ is semisimple, with a direct sum decomposition into eigenspaces

$$\mathfrak{g} = \sum_{\alpha} \mathfrak{g}_\alpha,$$

where α ranges over certain characters of $\mathfrak{a}$. In fact, we have a direct sum decomposition

$$\mathfrak{g} = \mathfrak{a} + \mathbf{i}\mathfrak{a} + \mathfrak{n} + {}^t\mathfrak{n}.$$

Then $\mathfrak{a} + \mathbf{i}\mathfrak{a} = \mathfrak{a} + \mathfrak{t}$ is the 0-eigenspace, and $\mathfrak{n}$ has the semisimple decomposition

$$\mathfrak{n} = \sum_{\alpha \in \mathcal{R}(\mathfrak{n})} \mathfrak{n}_\alpha,$$

taken over certain non-zero characters of $\mathfrak{a}$, which we call the **$(\mathfrak{a}, \mathfrak{n})$-characters**, or also the **$\mathfrak{n}$-relevant** characters. The set of these characters is denoted by $\mathcal{R}(\mathfrak{n})$.

Let $E_{ij} (i < j)$ be the matrix with (i, j)-component equal to 1, and all other components equal to 0. Then the elements

$$\{E_{ij}, \mathbf{i}E_{ij}\}$$

form a real basis for the eigenspace with eigencharacter α_{ij} such that if $H \in \mathfrak{a}$, H is a diagonal matrix, $H = \mathrm{diag}(h_1, \ldots, h_n)$, then

$$\alpha_{ij}(H) = h_i - h_j.$$

We write E_α instead of E_{ij} if α is the character α_{ij}.

The **simple** (eigen) **characters** are defined to be the eigencharacters

$$\alpha_i = \alpha_{i,i+1} \quad \text{with} \quad i = 1, \ldots, n-1.$$

The set of simple $(\mathfrak{a}, \mathfrak{n})$-characters is denoted by $\mathcal{S}(\mathfrak{n})$. Observe that an arbitrary eigencharacter is a sum of simple characters, that is

$$\alpha_{ij} = \alpha_{i,i+1} + \ldots + \alpha_{j-1,j}.$$

Thus the simple characters generate all relevant characters.

Going back to the decomposition $\mathfrak{g} = \mathfrak{a} + \mathbf{i}\mathfrak{a} + \mathfrak{n} + {}^t\mathfrak{n}$, letting $\mathcal{R}({}^t\mathfrak{n})$ denote the ${}^t\mathfrak{n}$-relevant characters (occurring in the semisimple decomposition of ${}^t\mathfrak{n}$ over $\mathfrak{a}$), we have trivially

$$\mathcal{R}({}^t\mathfrak{n}) = -\mathcal{R}(\mathfrak{n}).$$

In other words, the eigencharacters of ${}^t\mathfrak{n}$ are precisely the characters $-\alpha$ with $\alpha \in \mathcal{R}(\mathfrak{n})$. Instead of ${}^t\mathfrak{n}$ we therefore also write $\mathfrak{n}^-$.

The **trace** of the regular representation of $\mathfrak{a}$ on $\mathfrak{n}$ will be denoted by τ, so

$$\tau = \text{tr}\,(\mathcal{R}(\mathfrak{n})) = \sum_{\alpha \in \mathcal{R}(\mathfrak{n})} m(\alpha)\alpha,$$

where $m(\alpha) = \dim\ \mathfrak{n}_\alpha = 2$ on $\mathrm{SL}_n(\mathbf{C})$. More frequently, we use the half trace

$$\rho_G = \rho = \frac{1}{2}\text{tr}(\mathcal{R}(\mathfrak{n})) = \frac{1}{2} \sum_{\alpha \in \mathcal{R}(\mathfrak{n})} m(\alpha)\alpha.$$

Since $m(\alpha) = 2$ for the complex group G, one may also be led to deal with

$$\rho_0 = \frac{1}{2}\rho_G = \frac{1}{2} \sum_{\alpha \in \mathcal{R}(\mathfrak{n})} \alpha.$$

The simple characters $\{\alpha_1, \ldots, \alpha_r\}$ form a basis of $\mathfrak{a}^\vee$ over $\mathbf{R}$. Let $\{\alpha_1', \ldots, \alpha_r'\}$ be the dual basis. For a diagonal matrix $H = \text{diag}(h_1, \ldots, h_n)$ we have

$$\alpha_i'(H) = h_1 + \ldots + h_i, \text{ also denoted by } \lambda_i(H).$$

Furthermore,

$$\rho_0 = \lambda_1 + \ldots + \lambda_{n-1} \quad \text{and so} \quad \rho_G = 2(\lambda_1 + \ldots + \lambda_{n-1}).$$

The real trace form. On $\mathfrak{g}$ we have a non-singular $\mathbf{c}(G)$-invariant symmetric bilinear form, which is the real trace form, defined by

$$\langle Z, Z' \rangle = \text{Re tr}(ZZ').$$

It induces the preceding positive definite scalar product on $\mathfrak{a}$. Note that with respect to this form, two eigenspaces $\mathfrak{n}_\alpha, \mathfrak{n}_\beta$ are orthogonal for $\alpha, \beta \in \mathcal{R}(\mathfrak{n})$, unless $\alpha = \pm\beta$. This is done by direct verification in the present case. Thus we get an orthogonal decomposition

$$\mathfrak{g} = (\mathfrak{a} + \mathbf{i}\mathfrak{a}) + \sum_{\alpha \in \mathcal{R}(\mathfrak{n})} (\mathfrak{g}_\alpha + \mathfrak{g}_{-\alpha}) = (\mathfrak{a} + \mathbf{i}\mathfrak{a}) + \sum_{\alpha \in \mathcal{R}(\mathfrak{n})} (\mathfrak{n}_\alpha + {}^t(\mathfrak{n}_\alpha)).$$

In other words, aside from the above mentioned orthogonality, the space $\mathfrak{a} + \mathbf{i}\mathfrak{a}$ is orthogonal to each $\mathfrak{g}_\alpha$ and $\mathfrak{g}_{-\alpha}$.

Positivity. An element $H \in \mathfrak{a}$ is called **positive** if $\alpha(H) > 0$ for all the simple characters $\alpha \in \mathcal{S}(\mathfrak{n})$ (or equivalently, for all $\alpha \in \mathcal{R}(\mathfrak{n})$). The cone of positive elements is denoted by $\mathfrak{a}^+$ or $\mathfrak{a}_{>0}$. Positivity defines a partial ordering on $\mathfrak{a}$. On $\mathfrak{a}$, we also have the trace form (bilinear, with the matrix trace), defined on two elements $H, H' \in \mathfrak{a}$ by $\mathrm{tr}(HH')$. This trace form is positive definite, making $\mathfrak{a}$ into a euclidean space, and giving an isomorphism of $\mathfrak{a}$ with its dual space $\mathfrak{a}^\vee$. Positivity on $\mathfrak{a}^\vee$ is then defined so that this isomorphism preserves positivity. *An element $\lambda \in \mathfrak{a}^\vee$ is positive if and only if in terms of the dual basis*

$$\lambda = \sum s_i \lambda_i$$

with coefficients $s_i > 0$. Thus for some purposes, the dual basis is natural.

We let $A^+ = \exp \mathfrak{a}^+$. Thus A^+ consists of those $a \in A$ such that $a^\alpha > 1$ for all $\alpha \in \mathcal{R}(\mathfrak{n})$. We let A' be the subset of A consisting of **regular elements**, meaning $a \in A$ such that $a^\alpha \neq 1$ for all $\alpha \in \mathcal{R}(\mathfrak{n})$, or equivalently, if $a = \mathrm{diag}(a_1, \dots, a_n)$ then the components $a_1, \dots a_n$ are distinct. The Weyl group W (group of permutations of the diagonal elements) acts on A and on A', and A^+ is a fundamental domain for this action on A'.

Haar measure. For $a \in A$, we let

$$\delta(a) = a^{2\rho}.$$

From the Iwasawa decomposition $G = UAK$, a Haar measure on G is given in terms of the Iwasawa coordinates by the formula

$$\int_G f(x) d\mu(x) = \int_U \int_A \int_K f(uak)\delta(a)^{-1} du da dk.$$

Cf. [**JoL 01a**], Chapter I, §2. Thus we call δ the **Iwasawa character**.

CHAPTER 1

Estimates on SL_n Parabolics

This first chapter describes subgroups of G more general than the subgroup UA coming from the Iwasawa decomposition $G = UAK$. These subgroups contain UA, and the world is made up so that they constitute a complete system which allows both what is called parabolic induction, and a complete spectral decomposition formula on G/K, modulo a discrete subgroup Γ on the left. Thus they contain all the information relevant for us about $\Gamma\backslash G$, in a neat package.

The continuous part of the spectral decomposition is determined by what is called Eisenstein series. The chapter also give basic estimates which will be used subsequently to determine regions of convergence for these series, and estimates for them.

The background book [**JoL 01a**] was written on $\mathrm{SL}_n(\mathbf{R})$. As pointed out in the last chapter, to make some things simpler, notably the heat kernel, we actually use $\mathrm{SL}_n(\mathbf{C})$. Since our purpose in the present series of books is to exhibit the power of the heat kernel, and to minimize other difficulties arising from Lie theory, differential geometry or algebraic geometry, we go to $\mathrm{SL}_n(\mathbf{C})$ right away in the present book.

1.1. The hermitian norm on SL_n and Siegel sets

For definiteness, let $G = \mathrm{SL}_n(\mathbf{C})$. For $x \in \mathrm{Mat}_n(\mathbf{C})$ we let

$$x^* = {}^t\bar{x} \qquad \text{and} \qquad \theta x = {}^t\bar{x}^{-1} \quad \text{for} \quad x \in \mathrm{GL}_n(\mathbf{C}).$$

Thus x is unitary if and only if $\theta x = x$. For $x \in \mathrm{GL}_n(\mathbf{C})$ we have the usual polar Cartan-Lie decomposition

$$x = pk \quad \text{with} \quad p = (xx^*)^{1/2} \text{ hermitian positive definite and } k \text{ unitary.}$$

We can define the positive definite **hermitian trace form** on $\mathrm{Mat}_n(\mathbf{C})$ by

$$\langle x, y\rangle = \mathrm{tr}(xy^*).$$

In terms of coordinates, $\mathrm{tr}(xy^*) = \sum x_{ij}\bar{y}_{ij}$. We note the adjointness properties

(1a) $$\langle zx, y\rangle = \langle x, z^*y\rangle \quad \text{and} \quad \langle xz, y\rangle = \langle x, yz^*\rangle$$

(1b) $$\langle x, y\rangle = \langle y^*, x^*\rangle = \overline{\langle y, x\rangle}.$$

We define the corresponding hermitian **trace norm** on $\mathrm{Mat}_n(\mathbf{C})$ by

$$\|x\| = (\mathrm{tr}(xx^*))^{1/2} \qquad \text{and} \qquad \|x\|^2 = \sum |x_{ij}|^2.$$

Trivially, we have

$$\|x\| = \|x^*\|.$$

If $a_1^2, \dots, a_n^2$ are the eigenvalues of xx^*, with $a_i \geqq 0$, then

$$\|x\|^2 = \sum a_i^2.$$

There is no need for a complex conjugate here because $a_i \geqq 0$ for all $i = 1, \dots, n$. Recalling the inequality

$$(a_1 \cdots a_n)^{1/n} \leqq \frac{1}{n}(a_1 + \dots + a_n)$$

we conclude that

$$(\det\ xx^*)^{1/n} \leqq \frac{\operatorname{tr}(xx^*)}{n}. \tag{2}$$

For $x \in G = \mathrm{SL}_n(\mathbf{C})$, we have four basic properties of the hermitian norm.

HN 1. $\|x\| \geqq \sqrt{n}$ for all $x \in G$, and $\|k\| = \sqrt{n}$ for $k \in K$.
HN 2. $\|xy\| \leqq \|x\|\ \|y\|$ for $x, y \in G$.
HN 3. There is a constant c such that $\|x^{-1}\| \leqq c\|x\|^{n-1}$.
HN 4. K-bi-invariance, namely for $k_1, k_2 \in K$ we have
$\|k_1 x k_2\| = \|x\|$.

The first property **HN 1** is immediate from (2). The second property **HN 2** is the Schwarz inequality for the hermitian trace form mentioned above. The third property **HN 3** is immediate from the expression of the inverse matrix in terms of minor determinants, which introduce the power. The sums are bounded by such powers times a constant depending only on n. The hypothesis that $x \in \mathrm{SL}_n$ is of course used. Finally, property **HN 4** is immediate from the definition of K.

The K-bi-invariance of **HN 4** shows that we can express $\|x\|$ in terms of the A-polar component in the polar decomposition of x. This decomposition is

$$x = k_1 a k_2 \quad \text{with } k_1, k_2 \in K \text{ and } a \in A.$$

Let $a = \operatorname{diag}(a_1, \dots, a_n)$. Note that from the unitary property of k_1, k_2 we get

$$xx^* = k_1 a^2 k_1^{-1} \qquad \text{so} \qquad \|x\|^2 = \sum a_i^2.$$

In the above decomposition, we let $a = \mathrm{Pol}_A(x)$ denote the **polar projection** on A, well defined up to the operation of the Weyl group (permutations of the coordinates).

We define the **Lie height** by

$$\sigma(x) = |\log a| = |\log \mathrm{pol}_A(x)|,$$

where the norm $|\log\ a|$ is the norm associated with the trace form on the Lie algebra of A, so if $a = \operatorname{diag}(a_1, \dots, a_n)$ then

$$\sigma(x)^2 = \sum (\log\ a_i)^2.$$

LEMMA 1.1. *We have*

$$\log \|x\| \gg\ll \sigma(x) + 1 \quad \textit{for} \quad x \in G.$$

PROOF. Let $a = \mathrm{diag}(a_1, \dots, a_n) = \mathrm{Pol}_A(x)$. Then

$$\log \|x\|^2 = \log \ \mathrm{tr}(xx^*) = \log(\sum a_i^2) \gg\ll \max_i |\log \ a_i| + O(1)$$

by using **HN 1**. The inequalities of the lemma then follow by using the fact that on a euclidean space, the sup norm is equivalent to the euclidean norm.

We do not reproduce the proof that the σ-function above satisfies the triangle inequality

$$\sigma(xy) \leqq \sigma(x) + \sigma(y) \qquad \text{for} \quad x, y \in G.$$

The property was originally proved by Cartan using differential geometric aspects of the situation (reproduced in [**GaV 88**], Proposition 4.6.11), and was proved by Harish-Chandra using a direct Lie exponential argument [**Har 66**], §7, Lemma 10. Cf. [**JoL 01a**], Chapter X, Proposition 1.1. In light of **HN 2** and Lemma 1.1, we don't really need the precise triangle inequality since we deal with estimates up to a constant factor.

Orders of growth

A function f on G will be said to have **(at most) hermitian polynomial** growth if there exists $m > 0$ such that

$$f(x) = O(\|x\|^m) \qquad \text{for} \qquad \|x\| \to \infty.$$

A function f on G will be said to have **(at most) Lie polynomial growth** if there exists $m > 0$ such that

$$f(x) = O(1 + \sigma(x))^m \qquad \text{for} \qquad x \in G.$$

The function will be said to have **(at most) Lie exponential growth** if there exists a constant $c > 0$ such that

$$f(x) = O(e^{c\sigma(x)}) \quad \text{for} \quad x \in G.$$

From Lemma 1.1, we see that

Hermitian polynomial growth is equivalent to Lie exponential growth.

The function will be said to have **(at most) Lie exponential square growth** if

$$f(x) = O(e^{c\sigma^2(x)}) \text{ for some } c > 0 \text{ and all } x \in G.$$

We define **Lie polynomial decay, exponential decay**, and **exponential square decay** or **quadratic exponential decay** in the same way, but replacing the positive exponents by their negatives. For the significance of exponential square decay, see the Gaussian in Chapter 2, §2 and the Gauss space in Chapter 2, §3.

Next, we want to give an estimate of $\|x\|$ in terms of its Iwasawa A-component. We shall do this in Lemma 1.4 when x is restricted to Siegel sets, which we now define.

Let $\mathfrak{o} = \mathbf{Z}[\mathbf{i}]$. Let $\Gamma = \mathrm{SL}_n(\mathfrak{o})$. (Note that in the analogue case of $\mathrm{SL}_n(\mathbf{R})$, we let $\mathfrak{o} = \mathbf{Z}$ and $\Gamma = \mathrm{SL}_n(\mathbf{Z})$.) Let

$$G = UAK \qquad \text{and} \qquad \mathrm{Lie}(G) = \mathfrak{g} = \mathfrak{n} + \mathfrak{a} + \mathfrak{k}$$

be the Iwasawa decomposition. We write an element $x \in G$ in terms of its Iwasawa coordinates

$$x = uak, \qquad \text{so} \qquad a = x_A = \mathrm{Iw}_A(x).$$

We want to analyze $\Gamma\backslash G$, and we start with $\Gamma_U\backslash U$, where $\Gamma_U = \Gamma \cap U$. For $c > 0$, let:

U_c = set of elements $u \in U, u = I + Z, Z \in \mathfrak{n}$, and

$$|z_{ij}| \leqq c \quad \text{for all } (i,j) \text{ with } i < j.$$

Thus the z_{ij} are complex coordinates for elements of U, and U_c is compact. Note that $\Gamma_U = U(\mathfrak{o})$ is the set of elements in U having components in $\mathfrak{o}$.

LEMMA 1.2. *We have* $U = \Gamma_U U_{1/\sqrt{2}} = U_{1/\sqrt{2}}\Gamma_U$. *In particular,* $\Gamma_U\backslash U$ *is compact.*

PROOF. Since $\mathfrak{o} = \mathbf{Z}[\mathbf{i}]$ is a lattice in $\mathbf{C}$, with a square fundamental domain centered at 0, having sides of length 1, the above relation is proved by induction. We write $u = I + Z$ with Z strictly upper triangular, and adjust this element successively by multiplication with $I + W_j$ $(j = 2, \ldots, n)$ so as to move the j-th diagonal into the range where its components satisfy $|z_{ij}| \leqq 1/\sqrt{2}$. The procedure is the same as getting a square as fundamental domain for a torus, except that one has to take into account the non-commutativity of multiplication. Actually, the induction reduces the lemma to the abelian case because we can filter U by the subgroup U_j consisting of those elements $I + Z$ where Z has 0-components on the j'-diagonal, $2 \leqq j' < j$. Then U_j/U_{j+1} is actually a vector space, containing the lattice represented by elements with components in $\mathfrak{o}$, and we are in the torus situation.

Next, we have to deal with the A-coordinate. For $t > 0$, let:

A_t = subset of elements $a \in A$ such that $a^\alpha \geqq t$ for all simple $(\mathfrak{a}, \mathfrak{n})$-characters α.

If $a = \mathrm{diag}(a_1, \ldots, a_n)$ with $a_i > 0$ for $i = 1, \ldots, n$, then A_t consists of those elements $a \in A$ such that $a_i/a_{i+1} \geqq t$. Recall that the **simple characters** $\alpha_1, \ldots, \alpha_r$ are given by the formula

$$a^{\alpha_i} = a_i/a_{i+1}.$$

By a **Siegel set** $\mathfrak{S}$ we mean a subset of G of the form

$$\mathfrak{S}(\Omega_U, t) = \Omega_U A_t K$$

with a compact set Ω_U in U. If $\Omega_U = U_c$, then the Siegel set is also denoted by

$$\mathfrak{S}_{c,t} = U_c A_t K.$$

REMARK. *Let Ω be compact in U. Let* $\mathbf{c}$ *denote conjugation. Then*

$$\mathbf{c}(A_t^{-1})\Omega = \bigcup_{a \in A_t} a^{-1}\Omega a$$

so $\mathbf{c}(A_t^{-1})\Omega$ is bounded, hence relatively compact. This is immediate from the definitions, by using the semisimple decomposition of $\mathfrak{n}$ over $\mathfrak{a}$.

In general, there is no obvious relation between the size of the Iwasawa A-component of an element $x \in G$ and its polar A-component $\mathrm{Pol}_A(x)$. However, on a Siegel set, the two have essentially the same order of magnitude. Indeed, let

$$x = uak = k_1 b k_2$$

be the two decompositions. Then $ua^2\,{}^t\bar{u} = k_1 b^2 k_1^{-1}$. Taking the hermitian norm and using the fact that $\|u\|$ is bounded for x in a Siegel set, we get

$$\|a\|^2 \gg\ll \|b\|^2 \qquad \text{so} \qquad \|a\| \gg\ll \|b\|$$

or

$$\|\mathrm{Iw}_A(x)\| \gg\ll \|\mathrm{Pol}_A(x)\| \qquad \text{for } x \in \mathfrak{S}.$$

Siegel sets are used in the study of $\Gamma\backslash G$ as substitutes for fundamental domains, because they are easier to define and to handle. The reason why they can be used instead of fundamental domains lies in the following theorem.

THEOREM 1.3. *We have $G = \Gamma\mathfrak{S}_{c,t}$ for $t \leqq 1/\sqrt{2}$ and $c \geqq 1/\sqrt{2}$.*

Most of the rest of this section is devoted to the proof. It is put here for later convenience, but will not be used in the rest of the chapter. The proof amounts to a quantitative version of a euclidean algorithm. Qualitatively, one has in general: Let $\mathfrak{o}$ be a principal ring and F its quotient field. Let B_n be the Borel subgroup of upper triangular matrices in GL_n. Then $\mathrm{GL}_n(F) = \mathrm{SL}_n(\mathfrak{o})B_n(F)$.

The next results are classical, going under the name of reduction theory. We essentially follow Borel's exposition [**Bor 69**] for results of Hermite, Minkowski and Siegel. Theorem 1.3 implies that $\mathfrak{S}_{c,t}$ contains a fundamental domain of $\Gamma\backslash G$. For a more precise statement, especially a theorem of Grenier [**Gre 88**], [**Gre 93**], cf. [**JoL 01b**] Chapter I, Theorem 5.1. Actually, since we are dealing with $\mathrm{SL}_n(\mathbf{C})$ instead of $\mathrm{SL}_n(\mathbf{R})$, the constants in the theorem are slightly different from those in the above references.

A priori, we know that the sup norm of the coefficients of a matrix is equivalent to the hermitian norm. The first lemma shows that on a Siegel set, one may deal just with the first coefficient of the Iwasawa A-component.

We use the character λ_1 defined by $a^{\lambda_1} = a_1$ for $a = \mathrm{diag}(a_1, \dots a_n)$.

LEMMA 1.4.
(i) *Given* $t > 0$ *there exists* c_1 *such that if* $a \in A_t$ *then*

$$a_1^2 \leqq \|a\|^2 \leqq c_1 a_1^2.$$

(ii) *Given a Siegel set* $\mathfrak{S}$, *there exists* c_2 *such that for* $x \in \mathfrak{S}$, $x = uak$, *we have*

$$a_1 \ll \|x\| \leqq c_2 a_1.$$

(iii) *Given* $\lambda \in \mathfrak{a}^\vee$, *there exists* c_3 *and* N *such that for* $x \in \mathfrak{S}$,

$$x_A^\lambda \leqq c_3 \|x\|^N.$$

PROOF. By definition of A_t, we have $a_i^2/a_{i+1}^2 \geqq t^2$ for $i = 1, \ldots, r$, so $a_j^2 \ll a_1^2$ for all j. This proves the first inequality. For the second, we use property **HN 2**, which implies that for $x = uak$ in the Siegel set, we have

$$a_1 \leqq \|a\| = \|u^{-1}xk^{-1}\| \ll \|x\| \leqq \|u\|\|a\|\|k\| \ll \|a\|$$

because the hermitian norm is continuous and has a maximum on a compact set. We can then apply (i) to finish the proof of (ii). Conversely, given $\lambda \in \mathfrak{a}^\vee$, it is a linear combination of the projection characters on the diagonal coordinates. From this (iii) is immediate. The power N depends on the absolute value of the coefficients of the above linear combination.

Let $e_1, \ldots, e_n$ be the standard column unit vectors of $\mathbf{C}^n$. Their transposes ${}^te_1, \ldots, {}^te_n$ are then row vectors, and ${}^te_n = (0, \ldots, 0, 1)$. We use this vector for the following considerations instead of e_1 (as in Harish-Chandra or Borel) because we are using the Iwasawa decomposition, whereas they used the anti-Iwasawa decomposition. We are here following the practice on $\mathrm{SL}_2(\mathbf{R})$ from [**Lan 75/85**], and also on $\mathrm{Pos}_n(\mathbf{R})$ from [**JoL 01b**]. Let $\|\cdot\|$ also denote the hermitian norm on $\mathbf{C}^n$. We consider the function

$$g \mapsto \|{}^te_n g\| \quad \text{for} \quad g \in G.$$

REMARK. For $g \in G$, we have the Pythagorean expression

$$\|g\|^2 = \sum_{j=1}^n \|{}^te_j g\|^2.$$

The effect of using the above function is to eliminate the consideration of U and to concentrate on estimating the A-components. Indeed, we have ${}^te_n U = {}^te_n$, so U is contained in the isotropy group of ${}^te_n$ in G. Furthermore, ${}^te_n a = (0, \ldots, 0, a_n)$, and K acts unitarily. Hence if $g = uak$ is the Iwasawa decomposition,

$$\|{}^te_n g\| = \|{}^te_n uak\| = a_n.$$

Given $g \in G$, the function $\gamma \mapsto \|{}^te_n \gamma g\|$ has a minimum > 0 for $\gamma \in \Gamma$, because ${}^te_n \Gamma g$ is contained in the set of non-zero elements of a lattice in $\mathbf{C}^n$. For $g = \mathrm{id}$, it is the lattice of bottom rows with components in $\mathbf{Z}[\mathbf{i}]$. The next lemma provides the first step of an induction, carried out in the subsequent theorem.

LEMMA 1.5. *Let $g \in G$, and let $g = uak$ be its Iwasawa decomposition. Write $a = diag(a_1, \dots, a_n)$ with $a_i = a_i(g)$. Suppose that*

$$a_n = \|{}^t e_n g\| = \min_{\gamma \in \Gamma} \|{}^t e_n \gamma g\|.$$

Then

$$a_n \leqq \sqrt{2} a_{n-1} \qquad \text{that is} \qquad a^{\alpha_{n-1}} \geqq 1/\sqrt{2}.$$

PROOF. By Lemma 1.2 without loss of generality, we may assume that

$$|u_{n-1,n}|^2 \leqq 1/2.$$

Let $\gamma \in \Gamma$ permute the vectors ${}^t e_n$ and ${}^t e_{n-1}$ up to sign (to make γ have determinant 1). Then

$${}^t e_n \gamma g = \pm {}^t e_{n-1} g = \pm {}^t e_{n-1} uak = (0, \dots, 0, a_{n-1}, u_{n-1,n} a_n) k.$$

Hence

$$\begin{aligned} a_n^2 = \|{}^t e_n g\|^2 \leqq \|{}^t e_n \gamma g\|^2 &= a_{n-1}^2 + a_n^2 |u_{n,1,n}|^2 \\ &\leqq a_{n-1}^2 + \frac{1}{2} a_n^2. \end{aligned}$$

The lemma follows at once.

We now prove the theorem which implies Theorem 1.3. It gives a quantitative measure of the extent to which an orbit of Γ meets a Siegel set. Since the proof will be by induction on n, we have to put n in the notation for a special Siegel set, so we shall write

$$\mathfrak{S}^{(n)} = \mathfrak{S}^{(n)}_{1/\sqrt{2}, 1/\sqrt{2}}$$

for the Siegel set in $G_n = \mathrm{SL}_n(\mathbf{C})$ having the indicated constants $c = t = 1/\sqrt{2}$.

THEOREM 1.6. *Let $x \in G_n$. There exists $z \in \Gamma x \cap \mathfrak{S}^{(n)}$ such that*

$$\min_{\gamma \in \Gamma} \|{}^t e_n \gamma x\| = \|{}^t e_n x\|.$$

PROOF. Induction on n. There is nothing to prove for $n = 1$. Let $g \in \Gamma x$ be such that $\|{}^t e_n g\|$ is smallest, namely

$$\|{}^t e_n g\| \leqq \|{}^t e_n \gamma g\| \quad \text{for all } \gamma \in \Gamma,$$

so also

$$\|{}^t e_n g\| \leqq \|{}^t e_n \gamma x\| \quad \text{for all } \gamma \in \Gamma.$$

From an Iwasawa decomposition $g = uak$, we may write

$$g = \begin{pmatrix} g' a_n^{-1/(n-1)} & * \\ & \\ 0 \ \dots \ 0 & a_n \end{pmatrix} k \quad \text{so} \quad a_n = a_n(g),$$

with $g' \in \mathrm{GL}_{n-1}$. Then actually $g' \in \mathrm{SL}_{n-1}$ since $\det(g) = 1$. By induction, there exists $\gamma' \in \Gamma_{n-1}$ such that

$$\gamma' g' = \begin{pmatrix} a'_1 & & * \\ & \ddots & \\ 0 & & a'_{n-1} \end{pmatrix} k' \quad \text{with} \quad a'_i/a'_{i+1} \geqq 1/\sqrt{2}$$

for $i = 1, \dots, n-2$. Then

$$\begin{pmatrix} \gamma' & 0 \\ 0 & 1 \end{pmatrix} g = \begin{pmatrix} a'_1 a_n^{-1/(n-1)} & & & * \\ & \ddots & & \\ & & a'_{n-1} a_n^{-1/(n-1)} & \\ 0 & & & a_n \end{pmatrix} \begin{pmatrix} k' & 0 \\ 0 & 1 \end{pmatrix} k.$$

Let

$$y = \begin{pmatrix} \gamma' & 0 \\ 0 & 1 \end{pmatrix} g \quad \text{so} \quad a_n(y) = a_n(g) = a_n.$$

We have

$$\min_{\gamma \in \Gamma} \|{}^t e_n \gamma y\| = \min_{\gamma \in \Gamma} \|{}^t e_n \gamma g\| = a_n.$$

Note that $a_i(y) = a'_i a_n^{-1/n}$ for $i = 1, \dots, n-1$. Thus for $i = 1, \dots, n-2$,

$$\alpha_i(y) = a_i(y)/a_{i+1}(y) = a'_i/a'_{i+1} \geqq 1/\sqrt{2}.$$

By Lemma 1.5,

$$a_{n-1}(y)/a_n(y) \geqq 1/\sqrt{2}.$$

This takes care of the estimate for the A-component. Multiplying y on the left by some element in Γ_U does not change the A-component but changes y to an element z which lies in $\mathfrak{S}^{(n)}$ by using Lemma 1.2. This concludes the proof.

1.2. Volume and lattice point estimates

For concreteness, we let $G = \mathrm{SL}_n(\mathbf{C})$. For $t > 0$ we let:

$G(t)$ = set of all $x \in G$ such that $\|x\| \leqq t$.

Sometimes, it is convenient to abbreviate the notation, and let $G_t = G(t)$. One must then be sure that no confusion is possible with the notation for a Siegel set as in A_t. From the polar decomposition, we see at once that $G(t)$ is compact. Indeed, if $a = \mathrm{diag}(a_1, \dots, a_n)$ is the A-polar projection, then the $\|a_i\|$ are bounded, and so must also be bounded away from 0 (by **HN 1**), so the A-polar projection is compact.

Given a function $\psi(t)$ for $t \in \mathbf{R}_{>0}$, we say that ψ has **polynomial growth** if there exists $M > 0$ such that $\psi(t) = O(t^M)$ for $t \to \infty$.

We assume that the reader is acquainted with the standard Jacobian formulas and Haar measure computations, covered in [**JoL 01a**]. In particular, we assume familiarity with the Iwasawa character δ, which is the modular character on $\mathrm{SL}_n(\mathbf{C})$, as in Chapter I, §2 of that book. We use the character ρ such that

$$a^{2\rho} = \delta(a) \quad \text{so} \quad \rho = \frac{1}{2} \log \delta \circ \exp .$$

(Note that if δ_0 is the corresponding character on $\mathrm{SL}_n(\mathbf{R})$, then $\delta = \delta_0^2$.)

THEOREM 2.1. *The coset space $\Gamma\backslash G$ has finite (Haar) measure. A Siegel set in G has finite (Haar) measure.*

PROOF. By Theorem 1.3, $G = \Gamma\mathfrak{S}_{c,t}$ so the fact that a Siegel set has finite measure implies that $\Gamma\backslash G$ has finite measure. Since K is compact, and U_c has bounded (euclidean) measure, it follows from the Iwasawa coordinates integral formula that for some constant C,

$$\int\limits_{U_c A_t K} dg = C \int\limits_{A_t} \delta^{-1}(a) d^* a,$$

where d^*a is Haar measure on A. Hence it suffices to prove that this integral over A_t is finite. We use the coordinates $q_i = a_i/a_{i+1}$, $i = 1, \ldots , n-1$. Using the fact that Haar measure on each factor of $\mathbf{R}_{>0}^{(n-1)}$ is dq_i/q_i, and the fact that

$$\delta(a) = \prod_{i=1}^{n-1} q_i^{M_i}$$

with integers $M_i > 0$, we find that

$$\int\limits_{A_t} \delta^{-1}(a) d^* a = \int\limits_t^\infty \cdots \int\limits_t^\infty \prod q_i^{-M_i} \prod \frac{dq_i}{q_i}$$

which is finite, thus proving the theorem.

Next we shall use polar coordinates and the corresponding Jacobian. Let $m(\alpha)$ be the multiplicity of α ($= 2$ in the present case of $\mathrm{SL}_n(\mathbf{C})$), and let

$$J(a) = \prod_{\alpha \in \mathcal{R}(\mathfrak{n})} (a^\alpha - a^{-\alpha})^{m(\alpha)},$$

with the product taken over all the $(\mathfrak{a}, \mathfrak{n})$-characters. Then in polar coordinates,

$$\int_G f(x) dx = \int_K \int_{A^+} \int_K f(k_1 a k_2) J(a) dk_1 da dk_2 .$$

Cf. for instance [**JoL 01a**], Chapter VI, where it is done on $\mathrm{SL}_n(\mathbf{R})$, and the extension to $\mathrm{SL}_n(\mathbf{C})$ is similar. We now do [**Har 65**], Lemma 37. We let $A_{G(t)} = A \cap G(t)$.

LEMMA 2.2. *Let* vol *be the Haar volume on* G. *Then* $vol(G(t))$ *has polynomial growth for* $t \geqq 1$, *i.e. there exist* c, M *such that* $\mathrm{vol}(G(t)) \leqq ct^M$ *for* $t \geqq 1$.

PROOF. By the integration formula, using $\|x\| = \|\mathrm{Pol}_A(x)\|$, we get

$$\mathrm{vol}(G(t)) = \int_{A^+_{G(t)}} J(a)da \quad \text{up to a constant factor.}$$

For $a \in A^+$, we have

$$0 \leqq J(a) \leqq a^{2\rho} = e^{2\rho(\log\ a)}.$$

Since ρ is linear, we have $2\rho(\log\ a) \leqq M'|\log\ a|$ for some constant M', easily determined explicitly if one wants. Under the exponential map $a = \exp\ H$, the set $A^+_{G(t)}$ corresponds to a subset of $\mathfrak{a}^+(\log\ c_1 t)$, with some constant c_1, where $\mathfrak{a}^+(C)$ is the set of vectors H with $|H| \leqq C$. Hence the integral giving the volume $\mathrm{vol}(G(t))$ is bounded by

$$\int_{\mathfrak{a}^+(\log c_1 t)} t^{M'} dH,$$

from which the lemma follows.

Let $\mathfrak{o} = \mathbf{Z}[\mathbf{i}]$, and $\Gamma = \mathrm{SL}_n(\mathfrak{o})$. Let $\Gamma_{G(t)} = \Gamma \cap G(t)$. Let # denote the cardinality of a set. For each $t > 0$, $\#(\Gamma_{G(t)})$ is finite, immediately from the definitions.

LEMMA 2.3. *With the same* M *as in Lemma 2.2, we have*

$$\#(\Gamma_{G(t)}) = O(t^M) \quad \textit{for} \quad t \to \infty.$$

PROOF. Fix $G(B)$ with some $B > 0$. Pick B so small that no $\gamma \in \Gamma$ lies in $G(B)$ if $\gamma \neq \mathrm{id}$. Then for all $\gamma \in \Gamma$, γ is the only element of Γ in the translate $\gamma G(B)$. By the inequality **HN 2** of §1, $G(B)G(t) \subset G(Bt)$. Thus the volume of the union of the balls translated to the points of $\Gamma_{G(t)}$ grows at most like the volume $\mathrm{vol}(G_t)$, and we can apply Lemma 2.2 to conclude the proof.

The previous lemma is going to be applied to show the convergence of certain series, obtained by taking a Γ-trace, i.e. summing over all elements of Γ. We are doing [**Har 68**], Lemma 9 on SL_n, but under a condition weaker than compact support.

Let φ be a function on G. We say that φ has **polynomial decay** if there exists a positive integer N such that

$$|\varphi(z)| = O(\|z\|^{-N}) \quad \text{for} \quad \|z\| \to \infty.$$

Since the hermitian norm on G is bounded away from 0, this condition is equivalent to the existence of a constant C_N such that

$$|\varphi(z)| \leqq C_N \|z\|^{-N} \quad \text{for all} \quad z \in G.$$

If the condition holds, then we say that φ has polynomial decay **of order** N. At some point one uses functions with **superpolynomial decay**, meaning that the above estimate $|\varphi(z)| = O(\|z\|^{-N})$ for $\|z\| \to \infty$ holds for all N. We won't need this now.

LEMMA 2.4. *There exists N (depending only on G, Γ), and M' having the following property. Let φ be a function on G with polynomial decay of order N. Then there is a constant c such that*

$$\sum_{\gamma\in\Gamma} |\varphi(x\gamma y)| \leqq c\|x\|^{M'} \quad \textit{for all} \quad x, y \in G.$$

PROOF. Let $G(B-1, B)$ be the annulus consisting of all elements $x \in G$ such that

$$B - 1 \leqq \|x\| \leqq B.$$

Replacing φ be its absolute value we may assume $\varphi \geqq 0$. The sum in the lemma is dominated by

$$\sum_{B=1}^{\infty} \sum_{x\gamma y \in G(B-1,B)} \varphi(x\gamma y).$$

By **HN 3** in §1, for $z \in G(B)^{-1}$ we have $\|z\| \leqq c_1 B^m$ with some m. Furthermore,

$$x\gamma y \in G(B) \Leftrightarrow \gamma \in x^{-1}G(B)y^{-1}.$$

Let $\gamma_0 \in \Gamma \cap x^{-1}G(B)y^{-1}$. The map $\gamma \mapsto \gamma\gamma_0^{-1}$ gives an injection

$$\Gamma \cap x^{-1}G(B)y^{-1} \hookrightarrow x^{-1}G(B)G(B)^{-1}x.$$

Hence

$$\#\{\gamma \text{ such that } x\gamma y \in G(B)\} \leqq \#\{\gamma \text{ such that } \gamma \in x^{-1}G(B)G(B)^{-1}x\}.$$

By **HN 2** of §1, for γ in the set on the right, we have the inequality

$$\|\gamma\| \leqq c_1^2 \|x\|^{m+1} B^{m+1}$$

so by Lemma 2.3,

$$\#\{\gamma \text{ such that } x\gamma y \in G(B)\} \leqq c_2 c_1^{2M} (\|x\|B)^{M(m+1)}.$$

Thus we obtain

$$\sum_{x\gamma y \in G(B-1,B)} \varphi(x\gamma y) \ll c_2 c_1^{2M} (\|x\|B)^{M(m+1)} B^{-N}.$$

The lemma follows at once for N sufficiently large.

Define the operator T_φ by

$$(T_\varphi f)(x) = \int_G \varphi(xy) f(y) dy.$$

COROLLARY 2.5. *Let φ be a measurable function on G, with polynomial decay of order N sufficiently large. With M' and c as in Lemma 2.4, for every $f \in L^1(\Gamma\backslash G)$ and $x \in G$, we have*

$$|T_\varphi f(x)| \leqq c\|x\|^{M'}\|f\|_1.$$

PROOF. We get

$$\begin{aligned}|T_\varphi f(x)| &\leqq \int_G |\varphi(xy)f(y)|dy \\ &\leqq \int_{\Gamma\backslash G} \sum_{\gamma\in\Gamma} |\varphi(x\gamma y)f(y)|dy \leqq c\|x\|^{M'}\|f\|_1\end{aligned}$$

by Lemma 2.4, as desired.

1.3. Estimates of A-projections

This section contains estimates of A-projections under translations by Γ and under translations by elements in compact sets. We follow [**Har 68**], §1, adjusted to SL_n. The results of this section will be used in §6.

We continue with $G = \mathrm{SL}_n(\mathbf{C})$ and $\Gamma = \mathrm{SL}_n(\mathfrak{o})$. As usual, for $x \in G$ we let x_A be its A-projection in the Iwasawa decomposition. We recall that p_i $(i = 1, \ldots, r)$ is the multiplicative character defined by

$$p_i(a) = a_1 \ldots a_i.$$

Letting $\lambda_i = \log\ p_i$, we know that $\{\lambda_1, \ldots, \lambda_r\}$ is the dual basis $\{\alpha_1', \ldots \alpha_r'\}$ of $\{\alpha_1, \ldots, \alpha_r\}$. We have

$$\lambda_i(\log\ x_A) = \log\ p_i(x_A) \quad \text{or also} \quad p_i(x_A) = x_A^{\lambda_i}.$$

We use the alternating product representation of G on $\bigwedge^i \mathbf{C}^n$. We use $\|z\|$ to denote the euclidean norm of an element $z \in \mathbf{C}^n$, and we let $\{e_1, \ldots, e_n\}$ be the standard basis of unit vectors. The hermitian positive definite scalar product on $\bigwedge^d \mathbf{C}^n$ is chosen so that the elements

$$e_{j_1} \wedge \ldots \wedge e_{j_d} \quad \text{with} \quad j_1 < \ldots < j_d$$

form an orthonormal basis. For $k \in K$, the wedge product $\bigwedge^d k$ is unitary on $\bigwedge^d \mathbf{C}^n$.

For $u \in U$, we have

$$(\textstyle\bigwedge^d u)e_1 \wedge \ldots \wedge e_d = ue_1 \wedge \ldots \wedge ue_d = e_1 \wedge \ldots \wedge e_d.$$

For $a \in A$, we have

$$\textstyle\bigwedge^d(a)e_1 \wedge \ldots \wedge e_d = ae_1 \wedge \ldots \wedge ae_d = p_d(a)e_1 \wedge \ldots \wedge e_d.$$

LEMMA 3.1. *For all $\gamma \in \Gamma$, and $i = 1, \dots, r$, we have*

$$\gamma_A^{\lambda_i} = p_i(\gamma_A) \leqq 1 \quad or \quad \lambda_i(\log\ \gamma_A) \leqq 0.$$

PROOF. Non-zero elements of $\mathfrak{o}$ have absolute value $\geqq 1$. Let $\gamma = uak$ be the Iwasawa decomposition. Then

$$\begin{aligned} 1 &\leqq \| \textstyle\bigwedge^i(\gamma^{-1})e_1 \wedge \dots \wedge e_i\| = \|k^{-1}a^{-1}u^{-1}e_1 \wedge \dots \wedge k^{-1}a^{-1}u^{-1}e_i\| \\ &= \|a_1^{-1} \dots a_i^{-1} \textstyle\bigwedge^i(k^{-1})e_1 \wedge \dots \wedge e_i\| \\ &= p_i(a)^{-1}. \end{aligned}$$

This proves the lemma.

Next we return to Siegel sets, and do [**Har 68**], Chapter II, Lemma 21. For the general technique on SL_n, cf. [**JoL 01a**], Chapter I, §4 and Chapter V, §5.

LEMMA 3.2. *Let $\mathfrak{S}$ be a Siegel set. Let Ω be a compact set in G. Then uniformly for $x \in \mathfrak{S}$ and $y_1, y_2 \in \Omega$, we have*

$$(y_1xy_2)_A^{\lambda_i} = p_i((y_1xy_2)_A) \ll p_i(x_A) = x_A^{\lambda_i},$$

where the implied constant depends only on Ω and $\mathfrak{S}$.

PROOF. First, we note that

$$\| \textstyle\bigwedge^i(y_2^{-1}x^{-1}y_1^{-1})e_1 \wedge \dots \wedge e_i\| = p_i((y_1xy_2)_A^{-1})$$
$$\| \textstyle\bigwedge^i(x^{-1})e_1 \wedge \dots \wedge e_i\| = p_i(x_A)^{-1}.$$

Let $y_1^{-1} = k_1^{-1}b^{-1}v^{-1}$ with $y_1 = vbk_1$ the Iwasawa decomposition. We have

$$\begin{aligned} \| \textstyle\bigwedge^i(y_2^{-1}x^{-1}y_1^{-1})e_1 \wedge \dots \wedge e_i\| &= \| \textstyle\bigwedge^i(y_2^{-1}) \textstyle\bigwedge^i(x^{-1}y_1^{-1})e_1 \wedge \dots \wedge e_i\| \\ &\gg \| \textstyle\bigwedge^i(x^{-1}y_1^{-1})e_1 \wedge \dots \wedge e_i\| \end{aligned} \tag{1}$$

because $\bigwedge^i(y_2)$ is a bounded invertible operator, depending continuously on $y_2 \in \Omega$ and we can use

$$\inf_{y_2 \in \Omega} |\textstyle\bigwedge^i(y_2)|$$

as the implied constant in the inequality. Then

$$\begin{aligned} \| \textstyle\bigwedge^i(x^{-1}y_1^{-1})e_1 \wedge \dots \wedge e_i\| &= \| \textstyle\bigwedge^i(x^{-1}k_1^{-1}) \textstyle\bigwedge^i(b^{-1}v^{-1})e_1 \wedge \dots \wedge e_i\| \\ &= \| \textstyle\bigwedge^i(x^{-1}k_1^{-1})p_i(b^{-1})e_1 \wedge \dots \wedge e_i\|. \end{aligned}$$

Since y_1 ranges over a compact set, so does b^{-1}, the scalar factors $p_i(b^{-1})$ are bounded for $y_1 \in \Omega$, and the right side of (1) is

$$\gg \| \textstyle\bigwedge^i(x^{-1}k_1^{-1})e_1 \wedge \dots \wedge e_i\|. \tag{2}$$

Let $x = uak$ be the Iwasawa decomposition, so $x^{-1} = k^{-1}a^{-1}u^{-1}$. Then there is a compact set Ω_U such that for all $x \in \mathfrak{S}$,

$$x^{-1}k_1^{-1} = k^{-1}a^{-1}u^{-1}aa^{-1}k_1^{-1} \in K\Omega_U a^{-1}k_1^{-1}.$$

Hence the right side of (2) is

$$(3) \qquad \gg \| \bigwedge^i (a^{-1}k_1^{-1})e_1 \wedge \ldots \wedge e_i \|.$$

Since $\bigwedge^i(K)$ consists of unitary operators, we can write

$$\bigwedge^i(k_1^{-1}) = \sum_{j_1 < \ldots < j_i} c_{(j)} e_{j_1} \wedge \ldots \wedge e_{j_i}$$

with coefficients $c_{(j)}$ such that $\sum |c_{(j)}|^2 = 1$. Then

$$\bigwedge^i(a^{-1})\Lambda^i(k_1^{-1})e_1 \wedge \ldots \wedge e_i = \sum c_{(j)}p_{(j)}(a^{-1})e_{j_1} \wedge \ldots \wedge e_{j_i}$$

and

$$\| \bigwedge^i(a^{-1}k_1^{-1})e_1 \wedge \ldots \wedge e_i\|^2 = \sum |c_{(j)}|^2 p_{(j)}(a)^{-2}$$

$$(4) \qquad \gg \sum |c_{(j)}|^2 p_i(a)^{-2}$$

because $x \in \mathfrak{S}$, so $a \in A_t$, and $a_1 \ldots a_i \gg a_{j_1} \ldots a_{j_i}$, that is $p_{(j)}(a) \ll p_i(a)$. But then the right side of (4) is just $\gg p_i(a)^{-2}$, and Lemma 3.2 is proved.

Let $\lambda \in \mathfrak{a}^\vee$. Recall that $\lambda \geqq 0$ means either one of the following equivalent conditions (cf. [**JoL 01a**], Chapter I, §4).

We have $\langle \lambda, \alpha_i \rangle \geqq 0$ for all the simple $(\mathfrak{a}, \mathfrak{n})$-characters α_i $(i = 1, \ldots, r)$.

We can write λ as a linear combination

$$\lambda = \sum s_i \lambda_i \quad \text{with} \quad s_i \geqq 0,$$

where $\{\lambda_1, \ldots, \lambda_r\}$ is the dual basis of $\{\alpha_1, \ldots, \alpha_r\}$.

We then obtain a corollary of Lemmas 3.1 and 3.2.

LEMMA 3.3. (i) *Let* $\lambda \in \mathfrak{a}^\vee$, $\lambda \geqq 0$, *and* $\gamma \in \Gamma$. *Then*

$$\gamma_A^\lambda \leqq 1.$$

(ii) *Let* $\mathfrak{S}$ *be a Siegel set and* Ω *a compact set in* G. *Let* $\lambda \geqq 0$. *There exists* $c > 0$ *such that*

$$(y_1 x y_2)_A^\lambda \ll x_A^\lambda \quad \text{or} \quad \lambda((y_1 x y_2)_\mathfrak{a}) \leqq \lambda(x_\mathfrak{a}) + c$$

for all $x \in \mathfrak{S}$ *and* $y_1, y_2 \in \Omega$.

(iii) *With the same assumptions, there exists* $c_1 > 0$ *such that*

$$(\gamma y_1 x y_2)_A^\lambda \leqq c_1 x_A^\lambda$$

for all $x \in \mathfrak{S}$, $y_1, y_2 \in \Omega$ *and* $\gamma \in \Gamma$.

PROOF. Statements (i), (ii) are immediate from the second characterizing property of semipositivity, by applying Lemmas 3.1 and 3.2 respectively. For (iii), let $\gamma = u_\gamma a_\gamma k_\gamma$ be the Iwasawa decomposition of γ, so that $a_\gamma = \gamma_A$. Then

$$(\gamma y_1 x y_2)_A = a_\gamma (k_\gamma y_1 x y_2)_A.$$

We apply λ and use (i), (ii) with Ω replaced by $K\Omega$ to conclude the proof.

REMARK. The inequality $\gamma_A^\lambda \leqq 1$ with 1 on the right is special to the present choice of group. In general, one has to replace 1 by some positive constant, depending on the groups G and Γ. However, the essential feature of the above inequalities is that the effect of multiplication by $\gamma \in \Gamma$ on the left is to decrease the values of the positive characters, whether in their multiplicative form so they tend to 0 as γ goes to infinity, or their additive form so they tend to $-\infty$ as γ goes to infinity. A precise version of this rather loose statement will be given when we deal with Eisenstein series and give their precise domain of convergence, which requires the above lemma.

For convenience, we recall:

LEMMA 3.4. *Let $\lambda = s_1\lambda_1 + \ldots + s_r\lambda_r$ with $s_i > 0$ for all i. Then there exist constants c_1, c_2 such that for all $H \geqq 0$ (i.e. $\alpha_i(H) \geqq 0$ for all i) we have*

$$c_1|H| \leqq \lambda(H) \leqq c_2|H|.$$

PROOF. The right inequality just expresses the continuity of a functional. The left one is proved in the same way one proves two norms are equivalent. Let S be the unit sphere in $\mathfrak{a}$, i.e. the set of $H \in \mathfrak{a}$ such that $|H| = 1 = \langle H, H\rangle$. Then λ has a minimum at a point H in the compact set of elements $\geqq 0$ in S, and this minimum cannot be 0, otherwise $\lambda_i(H) = 0$ for all i so $H = 0$. The existence of c_1 follows by homogeneity.

1.4. Standard reduced parabolics

Given an integer $n \geqq 2$ we let $\mathcal{P}$ denote a partition of n, that is

$$n = n_1 + \ldots + n_{r+1} \quad \text{letting } r = r_{\mathcal{P}}$$

with positive integers $n_1, \ldots, n_{r+1}$. If $r = n-1, n = r+1$ we deal with the maximal partition. The partition is to be viewed as ordered, i.e. the sequence of integers $n_1, \ldots, n_{r+1}$ is given in this order. Equivalently, one could give the integers

$$m_i = n_1 + \ldots + n_i \quad \text{with} \quad i = 1, \ldots, r,$$

so that $1 \leqq m_1 < m_2 < \ldots < m_{r+1} = n$.

We consider blocks of $n_i \times n_i$ matrices along the diagonal with indices $i = 1, \ldots, r+1$. We let $G_n = \mathrm{SL}_n(\mathbf{C})$. Then:

$U_{\mathcal{P}} =$ subgroup of the unipotent upper triangular group with non-zero elements strictly above the square blocks, except for the diagonal elements equal to 1.

$A_{\mathcal{P}} =$ subgroup of the diagonal group A with positive diagonal elements which are constant in each block, the whole matrix having determinant 1.

$$G_{\mathcal{P}} = \prod_{i=1}^{r+1} \mathrm{SL}_{n_i} = \prod_{i=1}^{r+1} G_{n_i} = \text{direct product of the block groups.}$$

$K = K_n =$ unitary group SU_n.

$$K_{\mathcal{P}} = \prod_{i=1}^{r+1} K_{n_i} = \text{unitary subgroup of } G_{\mathcal{P}}.$$

The matrices in the above mentioned groups look as follows.

$$U_{\mathcal{P}} : \begin{bmatrix} \boxed{I_{n_1}} & & & * \\ & \boxed{I_{n_2}} & & \\ & & \ddots & \\ O & & & \boxed{I_{n_{r+1}}} \end{bmatrix}$$

$$G_{\mathcal{P}} : \begin{bmatrix} \boxed{G_{n_1}} & & & \\ & \boxed{G_{n_2}} & & O \\ & & \ddots & \\ O & & & \boxed{G_{n_{r+1}}} \end{bmatrix}$$

$$A_{\mathcal{P}} : \begin{bmatrix} \boxed{a_1 I_{n_1}} & & & \\ & \boxed{a_2 I_{n_2}} & & O \\ & & \ddots & \\ O & & & \boxed{a_{r+1} I_{n_{r+1}}} \end{bmatrix}$$

The components a_j $(j = 1, \dots , r+1)$ are subject to the determinant condition

$$\prod_{j=1}^{r+1} a_j^{n_j} = 1.$$

We define a **standard reduced parabolic** subgroup of G to be a subgroup of the form

$$P = U_{\mathcal{P}} A_{\mathcal{P}} G_{\mathcal{P}} \quad \text{also written} \quad P = U_P A_P G_P.$$

This is a subgroup. Indeed, G_P and A_P centralize each other. Furthermore, A_P and G_P normalize U_P, so P is a subgroup of G.

Note that G_P is an algebraic group, and a complex group in the complex case, i.e. when $G = \mathrm{SL}_n(\mathbf{C})$. Furthermore, P is a **maximal reduced parabolic** $(\neq G)$ if and only if G_P consists of two blocks:

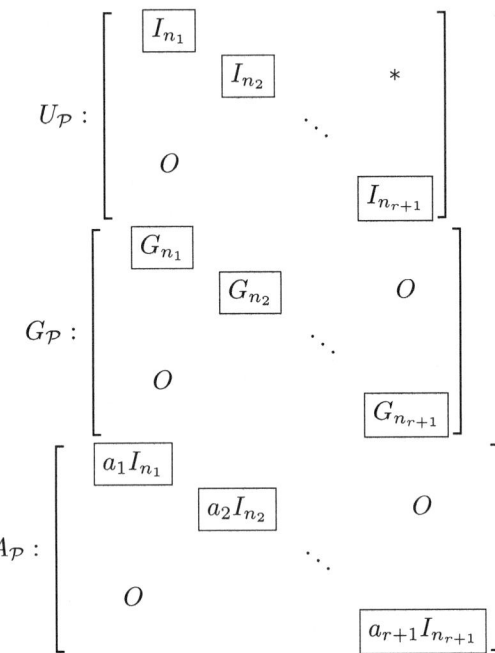

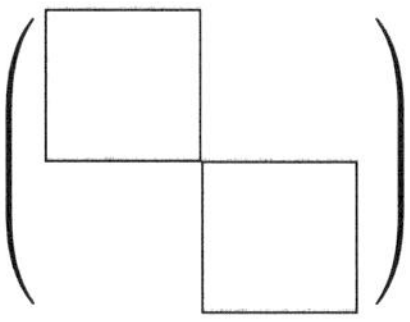

We define the groups:

$T_P =$ subgroup of diagonal matrices whose components have absolute value 1 and constant in each block.

$D_P = A_P T_P$.

$\tilde{G}_P = G_P T_P$.

Thus D_P is a diagonalized algebraic group, also called an **algebraic linear torus**. Over the complex numbers, it is isomorphic to an r_P-fold product of the multiplicative group. The algebraicity will be used later, when we shall need farther reaching foundational results from the theory of linear algebraic groups.

Then we also define the **standard parabolic** corresponding to P to be

$$\tilde{P} = U_P A_P T_P G_P = U_P A_P \tilde{G}_P \quad \text{(direct product decomposition)}.$$

For us, a parabolic will be much less important than the reduced parabolic because we are dealing with G/K and G_P/K_{G_P} in connection with right K-invariant functions. This introduces a simplification of the general theory.

An element $p \in \tilde{P}$ can be written uniquely as a product $p = ua\tilde{g}$ with $u \in U_P, a \in A_P$ and $\tilde{g} \in \tilde{G}_P$. Note that $G_P \cap T_P$ consists of diagonal matrices with m_i-th roots of unity in the i-th block, so we don't quite get a direct product decomposition $P = U_P D_P G_P$. The expression $ua\tilde{g}$ is called the **parabolic decomposition of** p. For $p \in P$, we have the **reduced parabolic decomposition** $p = uag$, with $g \in G_P$.

The groups $U_P A_P$ resp. $U_P A_P T_P$ are characterized as the maximal connected solvable normal subgroups of P resp. $\tilde{P}$. This is immediately verified, because $\mathrm{SL}_n(\mathbf{C})$ is simple.

In any linear group, there is a maximal normal unipotent subgroup, which is called the **unipotent radical**. *In the present case, U_P is the unipotent radical both of P and of $\tilde{P}$.* This is immediate, because G_P is a product of SL_{n_j} groups, and each factor has no non-trivial normal unipotent subgroup. The unipotent radical is invariant under any group automorphism. Under a group isomorphism, the unipotent radical of one group is mapped to the unipotent radical of the other.

If we let $K_{\tilde{P}} = K \cap \tilde{P}$, then $K_{\tilde{P}} = T_P K_{G_P}$.

We have given direct definitions. A more intrinsic group theoretic characterization is that the $\tilde{P}$'s are the algebraic sugbroups of G containing the minimal parabolics UD where D is the diagonal subgroup.

The central property of the standard reduced parabolic subgroups is that they determine the non-compact part of the spectral theory on G/K by means of an inductive procedure known as parabolic induction, in a sense to be made precise much later. We start a long journey by giving **parabolic coordinates** on G/K.

THEOREM 4.1. *The product map*

$$U_p \times A_P \times G_P/K_{G_P} \to G/K \quad \textit{given by} \quad (u, a, gK_{G_P}) \mapsto uagK$$

is a differential isomorphism.

Before giving the proper part of the proof, we make some remarks and give intermediate lemmas. First note that we took the coset space G_P/K_{G_P}. In some applications we consider right K_P-invariant functions in terms of the parabolic components (U_P, A_P, G_P). Such functions correspond to right K-invariant functions on G under the isomorphism of the theorem.

Note that

$$A_P \subset \text{ centralizer of } G_P.$$

If $(u, a, g) \in (U_P, A_P, G_P)$ we then have $uag = uga$. Note that we also have the ordinary U-Iwasawa component of an element $g \in G_P$, that is

$$g = u_g a_g k_g \quad \text{with} \quad u_g \in U_{G_P}, a_g \in A_{G_P} \quad \text{and} \quad k_g \in K_{G_P}.$$

The group U_{G_P} consists of blocks of unipotents along the diagonal.

We let

$$G = UAK$$

be the standard Iwasawa decomposition of G. Thus U is the upper triangular unipotent group, A is the group of positive diagonal matrices, and K is the unitary subgroup of G. Note that $U \supset U_P$ and $A \supset A_P$ for all P.

LEMMA 4.2. *The product map*

$$U_P \times U_{G_P} \to U \quad \textit{given by} \quad (u, v) \mapsto uv$$

is a C^∞ isomorphism. (Actually, it is an algebraic isomorphism.)

PROOF. First we note the injectivity of the map. Suppose $uv = u'v'$ with $u, u' \in U_P$ having their non-diagonal components above the blocks, and $v, v' \in U_{G_P}$ having their non-diagonal components inside the blocks. If $uv = u'v'$ then

$$u^{-1}u' = v(v')^{-1},$$

so each side is equal to the unit matrix and $u = u', v = v'$, thus showing injectivity. For two arbitrary $n \times n$ matrices $g, h \in P$, the diagonal blocks of gh are obtained by taking the product of the diagonal blocks of g and those of h respectively, so given $u \in U$ we let u_{G_P} be the matrix obtained by replacing the components of u above the blocks by 0. Then $u_{G_P}^{-1}u \in U_P$, thus proving the surjectivity. Note that both the product map and the inverse are C^∞ (actually algebraic), so the lemma is clear.

We can do with A what we just did with U. We have the two groups A_P (constant inside each block) and

$$A_{G_P} = \prod_{i=1}^{r+1} A_{G_{n_i}}.$$

LEMMA 4.3. *The product map $A_P \times A_{G_P} \to A$ is an isomorphism.*

PROOF. Immediate.

From the two lemmas and the fact that A_P is contained in the centralizer of G_P, we get:

LEMMA 4.4. *The product map*

$$U_P \times A_P \times U_{G_P} \times A_{G_P} = U_P \times U_{G_P} \times A_P \times A_{G_P} \to UA$$

is a C^∞ isomorphism.

This takes care of the UA-part. There remains only to deal with the right K-component, but this follows at once from the uniqueness of the Iwasawa decomposition $G = UAK$. This concludes the proof of Theorem 4.1.

REMARK. Let $K_P = K \cap P$. Then

$$K_{G_P} = K_P.$$

Indeed, if $k \in K \cap P$ so $k = uag$ with $u \in U_P, a \in A_P, g \in G_P$, and $g = u'a'k'$ with $u' \in U_{G_P}, a' \in A_{G_P}$ and $k' \in K_{G_P}$, then $uau'a' \in UA$, so by Iwasawa on G we must have $k = k'$.

From Theorem 4.1 we can get an actual fourfold decomposition of G. For definiteness, suppose $G_n = \mathrm{SL}_n(\mathbf{C})$, with Lie algebra $\mathfrak{sl}_n(\mathbf{C})$. Let:

$\mathfrak{s}_n$ = subspace of hermitian elements, i.e. $Z \in \mathfrak{g}_n$ such that

$${}^tZ = \bar{Z} \quad \text{or equivalently} \quad \theta Z = -Z \quad \text{with} \quad \theta Z = -{}^t\bar{Z}.$$

Let $\mathbf{S}_n = \exp \mathfrak{s}_n$. Then $\mathbf{S}_n$ is not a subgroup but is a submanifold, consisting of the positive definite hermitian matrices. We have the direct product decomposition

$$G_n = \mathbf{S}_n K_n.$$

This decomposition applies to each one of the blocks in the expression of G_P as product, i.e., it applies to each G_{n_i} $(i = 1, \ldots, r+1)$. We let:

$$\mathbf{S}_P = \prod_{i=1}^{r+1} \mathbf{S}_{n_i} \quad \text{so} \quad G_P = \mathbf{S}_P K_{G_P} \quad \text{(direct product)}.$$

Then we obtain as a corollary of Theorem 4.1:

THEOREM 4.5. *The product map*

$$U_P \times A_P \times \mathbf{S}_P \times K \to G \quad \text{given by} \quad (u, a, x, k) \mapsto uaxk$$

is a C^∞ isomorphism.

NOTE:. For the general context and definitions, aside from Harish-Chandra's works, e.g. [**Har 75**], see for instance [**Var 77**], p.285, and [**GaV 88**], §2.3, especially 2.3.3.

We shall use Theorem 4.1 to extend functions from A_P and G_P to G as follows.

Concerning A_P, Theorem 4.1 yields a linear isomorphism.

$$C(A_P) \xrightarrow{\approx} C(U_P G_P \backslash G/K).$$

Going from G to A_P, the inverse isomorphism is the map $F_G \mapsto F_{A_P}$ obtained by restriction to A_P. Going from A_P to G, given a continuous function h on A_P, we can extend h to a function h_G on G such that $h_G \in C(U_P G_P \backslash G/K)$, so

$$h = (h_G)_{A_P},$$

and h_G is the composite of h and the projection on A_P. In other words,

$$h_G(x) = h(x_{A_P}) \quad \text{for } x \in G.$$

By x_{A_P} we mean the A_P-coordinate of x in the product expression of Theorem 4.1.

Concerning G_P, Theorem 4.1 yields a linear isomorphism

$$C(G_P/K_P) \xrightarrow{\approx} C(U_P A_P \backslash G/K).$$

Given F continuous on G_P/K_P, we can extend F uniquely to a function F_G on G with left $U_P A_P$-invariance and right K-invariance, so a function on G/K. Thus

$$F_G(x) = F(x_{G_P/K_{G_P}}) \quad \text{for } x \in G.$$

Decomposition of the discrete subgroups

We let $\Gamma = \Gamma_n = \mathrm{SL}_n(\mathbf{Z}[\mathbf{i}])$, formed with the ring of Gaussian integers. Then Γ is a discrete subgroup of $G = \mathrm{SL}_n(\mathbf{C})$, and the homogeneous space

$$\mathrm{SL}_n(\mathbf{Z}[\mathbf{i}]) \backslash \mathrm{SL}_n(\mathbf{C}) = \Gamma \backslash G$$

has finite volume for its Haar measure, i.e. the G-invariant measure induced from the action of $\mathrm{SL}_n(\mathbf{C})$.

Let H be a subgroup of G. We use the notation

$$\Gamma_H = \Gamma \cap H.$$

Thus we obtain the subgroups $\Gamma_P, \Gamma_{U_P}, \Gamma_{G_P}$. Note that Γ_{A_P} is the trivial group.

For a reduced standard parabolic subgroup P as above, we have the semidirect product decomposition

$$\Gamma_P = \Gamma_{U_P} \Gamma_{G_P}. \tag{1}$$

This statement is essentially trivial. Let $\gamma \in \Gamma_P$, and take $\gamma_1, \ldots, \gamma_{r+1}$ to be the block components of γ along the diagonal. Then the block diagonal matrix $\mathrm{diag}(\gamma_1, \ldots, \gamma_{r+1})$ is in $A_P G_P$ and cannot have an A_P-factor other than the identity. Indeed, let $\mathrm{diag}(a_1 I_{n_1}, \ldots, a_{r+1} I_{n_{r+1}})$ be its A_P-factor, and $\gamma_j = a_j g_j$ with $\mathrm{diag}(g_1, \ldots, g_{r+1}) \in G_P$, so $\det g_j = 1$ for all j. Then $a_j^{n_j}(\det g_j)$ has components in $\mathbf{Z}[\mathbf{i}]$, whence $a_j^{n_j} \in \mathbf{Z}[\mathbf{i}]$, so $a_j^{n_j} \in \mathbf{Z}$ because a_j is real positive. Since we also have $\prod a_j^{n_j} = 1$, it follows that $a_j = 1$ for all j, as stated. Hence $\gamma_j \in G_{n_j}$ for each j. Let

$$\gamma_{G_P} = \begin{pmatrix} \gamma_1 & & \ldots & 0 \\ \vdots & \gamma_2 & & \vdots \\ & & \ddots & \\ 0 & \ldots & & \gamma_{r+1} \end{pmatrix} \quad \text{and} \quad \gamma_{U_P} = \gamma_{G_P}^{-1}\gamma.$$

Then $\gamma_{U_P} \in \Gamma_{U_P}$ and we get the desired decomposition. The uniqueness is immediate.

Next, let $x \in P$. With respect to the reduced parabolic decomposition $P = U_P A_P G_P$, let x_{U_P}, x_{A_P} and x_{G_P} denote the components of x on the respective factors. Let $\gamma \in \Gamma_P$. Then

$$(2) \qquad (\gamma x)_{A_P} = x_{A_P} \quad \text{and} \quad (\gamma x)_{G_P} = \gamma_{G_P} x_{G_P}.$$

PROOF. Let $x = uag$ be its reduced parabolic decomposition. Then by (1),

$$\gamma x = \gamma_{U_P}\gamma_{G_P} uag = \gamma_{U_P}\gamma_{G_P} u \gamma_{G_P}^{-1} a \gamma_{G_P} g,$$

because g commutes with a. Hence a is the A_P-projection of both γx and x, thus proving the first formula. The second formula also follows from (1) and the above equality.

The group Γ_{U_P} is the subgroup of the unipotent group U_P such that the components strictly above the diagonal are in $\mathfrak{o}$. We have:

LEMMA 4.6. *The coset space $\Gamma_{U_P}\backslash U_P$ is compact.*

PROOF. In Lemma 1.2 we had already seen that $\Gamma_U\backslash U$ is compact, and $\Gamma_{U_P}\backslash U_P$ is a closed subspace of $\Gamma_U\backslash U$, so the assertion is clear.

REMARK. The spaces $\Gamma_{U_P}\backslash U_P$ and $\Gamma_U\backslash U$ behave like toruses, even though they are in general not commutative, or even groups.

1.5. Characters on the parabolics

We continue with the same notation, so with a partition of n and the resulting groups U_P, A_P, G_P, K_{G_P}. The groups $A_P, G_P, K_{G_P}, U_{G_P}$ are block groups of the same type as A, G, K, U respectively. We then have the Lie algebras

$$\mathfrak{n}_P = \text{Lie}(U_P), \quad \mathfrak{a}_P = \text{Lie}(A_P), \quad \mathfrak{g}_{G_P} = \text{Lie}(G_P), \quad \mathfrak{k}_{G_P} = \text{Lie}(K_{G_P}).$$

Corresponding to the Iwasawa decomposition $G = UAK$, we have:

$\mathcal{R}(\mathfrak{n})$ = set of $(\mathfrak{a}, \mathfrak{n})$-**characters**, also called the $\mathfrak{n}$-**relevant characters**.

These are the characters of $\mathfrak{a}$ occurring in the semisimple decomposition of $\mathfrak{n}$ over the Lie-regular action of $\mathfrak{a}$. We denote such characters by α. We let:

$\mathcal{S}(\mathfrak{n})$ = set of **simple characters** in $\mathcal{R}(\mathfrak{n})$-namely in the usual notation the characters

$$\alpha_i = \alpha_{i,i+1} \quad \text{with } i = 1, \dots, n-1.$$

Cf. [**JoL 01**], Chapter I and Chapter III, §2. Given an element $H \in \mathfrak{a}$,

$$H = \begin{pmatrix} h_1 & & 0 \\ & \ddots & \\ 0 & & h_n \end{pmatrix} \quad \text{with} \quad \text{tr}(H) = \sum h_i = 0,$$

by definition $\alpha_i(H) = h_i - h_{i+1}$. The relevant characters are sums of simple characters. On SL_n, as we are now, they are the characters α_{ij} with $i < j$, such that

$$\alpha_{ij}(H) = h_i - h_j \quad \text{so} \quad \alpha_{ij} = \alpha_i + \ldots + \alpha_{j-1}.$$

We use the real trace form on $\mathfrak{g}$ as the standard scalar product. For $\text{SL}_n(\mathbf{R})$, we thus have

$$B(X, X') = \langle X, X' \rangle = \text{tr}(XX').$$

For $\text{SL}_n(\mathbf{C})$,

$$B(Z, Z') = \langle Z, Z' \rangle = \text{Re tr}(ZZ').$$

The essential property is that this scalar product (bilinear form) is G-invariant. Furthermore, the bilinear form on the complex Lie algebra restricts to the bilinear form on the real Lie algebra. Some of the literature in the complex case takes 2 times the real trace form, so occasionally our normalization differs by a factor of 2 from the one found in some other references.

If $\alpha \in \mathfrak{a}^\vee$ (the dual space of $\mathfrak{a}$), then we let $H_\alpha \in \mathfrak{a}$ be the vector such that

$$\alpha(H) = \langle H_\alpha, H \rangle.$$

We shall describe the set of characters analogous to $\mathcal{R}(\mathfrak{n})$ and $\mathcal{S}(\mathfrak{n})$ as well as duality, for the parabolic case. We let:

$\mathcal{R}(\mathfrak{n}_{G_P})$ = subset of characters $\alpha \in \mathcal{R}(\mathfrak{n})$ such that $\alpha(\mathfrak{a}_P) = 0$ or equivalently $a^\alpha = 1$ for all $a \in A_P$.

Thus $\mathcal{R}(\mathfrak{n}_{G_P})$ consists of those eigencharacters in $\mathcal{R}(\mathfrak{n})$ which occur in the $\mathfrak{a}$-semisimple decomposition of $\mathfrak{n}_{G_P}$. In terms of the blocks of indices, we have the alternative description:

$\mathcal{R}(\mathfrak{n}_{G_P})$ = subset of $\mathcal{R}(\mathfrak{n})$ consisting of the characters α_{ij} with

$i < j$ such that i, j belong to a block of indices.

The α_{ij} with $i < j$ and i, j in different blocks are thus omitted. We let

$\mathcal{S}(\mathfrak{n}_{G_P})$ = set of simple characters in $\mathcal{S}(\mathfrak{n})$ which occur in the $\mathfrak{a}$-semisimple decomposition of $\mathfrak{n}_{G_P}$, or equivalently which are 0 on $\mathfrak{a}_P$.

In terms of the blocks, this gives:

$\mathcal{S}(\mathfrak{n}_{G_P})$ = subset of elements $\alpha \in \mathcal{S}(\mathfrak{n})$ arising from inside the blocks, that is precisely the elements

$$\{\alpha_1, \ldots, \alpha_{n_1-1}, \alpha_{n_1+1}, \ldots, \alpha_{n_1+n_2-1}, \ldots\}.$$

Thus we omit the characters $\alpha_i \in \mathcal{S}(\mathfrak{n})$ which straddle two blocks, that is i, $i+1$ belong to two successive blocks of indices. Similarly, using $\mathfrak{n}_P$ instead of $\mathfrak{n}_{G_P}$, we define the set of **simple $\mathfrak{n}_P$-relevant characters** to be:

$\mathcal{R}(\mathfrak{n}_P)$ = subset of elements $\alpha \in \mathcal{R}(\mathfrak{n})$ which occur in the $\mathfrak{a}$-semisimple decomposition

$$\mathfrak{n}_P = \bigoplus_{\alpha} \mathfrak{n}_{P,\alpha}.$$

i.e. such that the α-eigenspaces $\mathfrak{n}_{P,\alpha}$ are $\neq 0$. In terms of blocks, these are the characters $\alpha = \alpha_{ij}$ for which i, j lie in different blocks of the partition. We also have the simple characters:

$\mathcal{S}(\mathfrak{n}_P)$ = set of simple characters which occur in the $\mathfrak{a}$-semi-simple decomposition of $\mathfrak{n}_P$
= set of characters $\alpha_{i,i+1}$ such that $i, i+1$ lie in two successive blocks.

Note that $\mathcal{S}(\mathfrak{n}_P)$ is a basis of $\mathfrak{a}_P^{\vee}$. We have the disjoint union

(3) $$\mathcal{R}(\mathfrak{n}) = \mathcal{R}(\mathfrak{n}_{G_P}) \cup \mathcal{R}(\mathfrak{n}_P) \quad \text{so} \quad \mathcal{R}(\mathfrak{n}_P) = \mathcal{R}(\mathfrak{n}) - \mathcal{R}(\mathfrak{n}_{G_P}).$$

Directly from the definitions, we see that the decomposition (3) is an orthogonal decomposition. We also have the disjoint union

(4) $$\mathcal{S}(\mathfrak{n}) = \mathcal{S}(\mathfrak{n}_{G_P}) \cup \mathcal{S}(\mathfrak{n}_P).$$

The elements of $\mathcal{S}(\mathfrak{n}_P)$ can be indexed in the form

$$\mathcal{S}(\mathfrak{n}_P) = \{\alpha_{P,1}, \ldots, \alpha_{P,r_P}\}.$$

Just as in the case of G, we define the **P-trace**

$$\tau_P = \sum_{\alpha \in \mathcal{R}(\mathfrak{n}_P)} m(\alpha)\alpha \quad \text{and} \quad \delta_P(a) = e^{\tau_P(\log\ a)},$$

and the **half trace**

(5a) $$\rho_P = \frac{1}{2}\tau_P = \frac{1}{2} \sum_{\alpha \in \mathcal{R}(\mathfrak{n}_P)} m(\alpha)\alpha.$$

The character δ_P is called the P-**Iwasawa character**, and plays the same role as δ in the parabolic coordinates integration formula, given in §7 below.

Of course, we also have the element

$$\rho_{G_P} = \frac{1}{2}\tau_{G_P} = \frac{1}{2}\sum_{\alpha \in \mathcal{R}(\mathfrak{n}_{G_P})} m(\alpha)\alpha. \tag{5b}$$

Recall that $m(\alpha) = 2$ on $\mathrm{SL}_n(\mathbf{C})$. From the disjoint union (4), we get the decomposition

$$\rho_G = \rho_{G_P} + \rho_P.$$

Scalar product and positivity

Since $\mathfrak{a}_P$ is a subspace of $\mathfrak{a}$, the trace form on $\mathfrak{a}$ induces a positive definite scalar product on $\mathfrak{a}_P$, which we also call the trace form, and which we denote by the same symbol $\langle \cdot, \cdot \rangle$. We then have various possibilities for positivity. For instance a character $\lambda \in \mathfrak{a}_P^\vee$ is $\mathcal{S}(\mathfrak{n}_P)$**-positive** means that

$$\langle \lambda, \alpha \rangle > 0 \quad \text{for all} \quad \alpha \in \mathcal{S}(\mathfrak{n}_P).$$

If $\{\alpha'_{P,1}, \ldots, \alpha'_{P,r_P}\}$ is the dual basis of $\mathcal{S}(\mathfrak{n}_P)$, then we have the dual notion of λ being $\mathcal{S}(\mathfrak{n}_P)'$**-positive**, namely

$$\langle \lambda, \alpha'_{P,i} \rangle > 0 \quad \text{for all } i = 1, \ldots, r_P.$$

For a general discussion of positivity, cf. [**JoL 01a**], Chapter I, §4. We assume the reader is acquainted with this basic material. In particular, λ is $\mathcal{S}(\mathfrak{n}_P)$-positive if and only if λ is a linear combination of the dual basis

$$\lambda = \sum s_i \alpha'_{P,i}$$

with $s_i > 0$ for all i.

The elements of $\mathcal{S}(\mathfrak{n})$ form a basis of $\mathfrak{a}^\vee$. Cf. [**JoL 01**], Chapter I, §4. We write these elements as $\{\alpha_1, \ldots, \alpha_{n-1}\}$, and we use $\{\alpha'_1, \ldots, \alpha'_{n-1}\}$ for the dual basis. If $H = \mathrm{diag}(h_1, \ldots, h_n)$ then $\alpha'_i(H) = h_1 + \ldots + h_i$. The disjoint decompositions (3) and (4) give rise to a disjoint decomposition of $\mathcal{S}(\mathfrak{n})'$ in terms of the dual bases $\mathcal{S}(\mathfrak{n}_P)'$ and $\mathcal{S}(\mathfrak{n}_{G_P})'$. We also get a decomposition of ρ. On $\mathrm{SL}_n(\mathbf{C})$,

$$\rho_G = 2\sum_{i=1}^{n-1} \alpha'_i \quad \text{(the factor 2 comes from } m(\alpha) = 2). \tag{6}$$

In the real case, we get $\frac{1}{2}$ the right side. From the definition of the dual basis, we get

$$\langle \rho_G, \alpha \rangle = 2 \quad \text{for } \alpha \in \mathcal{S}(\mathfrak{n}), \tag{7}$$

and

$$\langle \rho_G, \alpha \rangle = 2 \quad \text{for } \alpha \in \mathcal{S}(\mathfrak{n}_{G_P}), \tag{7P}$$

In the real case, we get 1 on the right side.

LEMMA 5.1. *The decomposition*

$$\rho_G = \rho_{G_P} + \rho_P$$

is orthogonal, and more strongly,

$$\langle \rho_P, \alpha \rangle = 0 \quad \textit{for} \quad \alpha \in \mathcal{R}(\mathfrak{n}_{G_P}).$$

In particular,

$$\rho_G^2 = \rho_{G_P}^2 + \rho_P^2.$$

PROOF. This is immediate from (7) and (7P). Note that an element of $\mathcal{R}(\mathfrak{n}_{G_P})$ is a sum of elements in $\mathcal{S}(\mathfrak{n}_{G_P})$.

Next we give another interpretation of Lemma 5.1. By definition,

(8)
$$H_{\rho_G} = \frac{1}{2} \sum_{\alpha \in \mathcal{R}(\mathfrak{n})} m(\alpha) H_\alpha.$$

We also define

(8P)
$$H_{\tau_P} = \sum_{\alpha \in \mathcal{R}(\mathfrak{n}_P)} m(\alpha) H_\alpha \quad \text{and} \quad H_{\rho_P} = \frac{1}{2} H_{\tau_P}.$$

LEMMA 5.2. *With H_{τ_P} as above, we have $H_{\tau_P} \in \mathfrak{a}_P$.*

PROOF. From (7) and (7P) we get

$$\langle \rho_G - \rho_{G_P}, \alpha \rangle = 0 \quad \text{for} \quad \alpha \in \mathcal{R}(\mathfrak{n}_{G_P}).$$

By (8P) we conclude that $\alpha(H_{\tau_P}) = 0$ for all $\alpha \in \mathcal{R}(\mathfrak{n}_{G_P})$. Directly from the definition of $\mathcal{S}(\mathfrak{n}_{G_P})$ this implies that $H_{\tau_P} \in \mathfrak{a}_P$, namely in our concrete situation, that H_{τ_P} is constant in each block. This proves the lemma.

LEMMA 5.3. *The elements H_{ρ_P} and H_{τ_P} are $\mathcal{S}(\mathfrak{n}_P)$-positive.*

PROOF. The element $\rho = \rho_G$ is $\mathcal{S}(\mathfrak{n})$-positive because $\rho = 2\sum \alpha_i'$ is expressible as a positive linear combination of the dual basis $\mathcal{S}(\mathfrak{n})'$ of $\mathfrak{a}'^{\vee}$. The orthogonalities of Lemmas 5.1 and 5.2 conclude the proof.

1.6. Estimates of A_P-projections

We saw in Lemma 4.3 that

$$A = A_P A_{G_P} \quad \text{and} \quad \mathfrak{a} = \mathfrak{a}_P \oplus \mathfrak{a}_{G_P}.$$

We now formulate the analogues of Lemma 3.3 for the A_P-projection under the parabolic decomposition

$$G/K = U_P A_P G_P / K_P,$$

which we gave in Theorem 4.1. We follow Harish-Chandra as in §3.

LEMMA 6.1. *Let $\lambda \in \mathfrak{a}_P^\vee$ be $\mathcal{S}(\mathfrak{n}_P)$-semipositive, that is $\langle \lambda, \alpha \rangle \geqq 0$ for all $\alpha \in \mathcal{S}(\mathfrak{n}_P)$.*
(i) *For all $\gamma \in \Gamma$, we have*

$$\gamma_{A_P}^\lambda \leqq 1 \quad \text{or} \quad \lambda(\gamma_{\mathfrak{a}_P}) \leqq 0.$$

(ii) *Let $\mathfrak{S}$ be a Siegel set and Ω a compact set in G. Then*

$$(y_1 x y_2)_{A_P}^\lambda \ll x_{A_P}^\lambda \quad \text{for all } x \in \mathfrak{S} \text{ and } y_1, y_2 \in \Omega.$$

(iii) *There exists $c > 0$ such that*

$$(\gamma y_1 x y_2)_{A_P}^\lambda \leqq c x_{A_P}^\lambda \quad \text{for all } y_1, y_2 \in \Omega, x \in \mathfrak{S}, \gamma \in \Gamma.$$

PROOF. The first two statements are proved by direct reduction to the corresponding statements on A (with respect to the full Iwasawa decomposition). We extend λ to an element Λ of $\mathfrak{a}^\vee$ by giving it the value 0 on the direct summand $\mathfrak{a}_{G_P}$. Then

$$x_A^\Lambda = x_{A_P}^\lambda \quad \text{for all } x \in G.$$

Let $\alpha \in \mathcal{S}(\mathfrak{n})$ and let $\alpha_{\mathfrak{a}_P}$ be its restriction to $\mathfrak{a}_P$. Then

$$\langle \Lambda, \alpha \rangle = \langle \lambda, \alpha_{\mathfrak{a}_P} \rangle \geqq 0.$$

This establishes the desired reduction to Lemma 3.3(i), (ii).

As to (iii), let $\gamma = u_\gamma a_\gamma g_\gamma k_\gamma$ be a parabolic coordinate decomposition of γ, so $a_\gamma = \gamma_{A_P}$. Then

$$(\gamma y_1 x y_2)_{A_P} = a_\gamma (k_\gamma y_1 x y_2)_{A_P}.$$

We apply λ and use (i), (ii) with the compact set $K\Omega$ to conclude the proof.

We give what amounts to an alternative formulation of the above lemma. As in the previous section, let $\mathcal{S}(\mathfrak{n}_P)' = \{\alpha'_{P,1}, \ldots, \alpha'_{P,r_P}\}$ be the dual basis of the simple characters $\mathcal{S}(\mathfrak{n}_P)$. Let:

^-A_P = set of elements $a \in A_P$ such that $\alpha'_{P,i}(\log\ a) < 0$ for all $i = 1, \ldots, r_P$.

We call the elements of ^-A_P the $\mathcal{S}(\mathfrak{n}_P)'$**-log negative elements**. We give explicitly a convenient reformulation of Lemma 6.1(iii).

LEMMA 6.2. *Let Ω be a compact set in G and $\mathfrak{S}$ a Siegel set. Let P be a reduced standard parabolic. Then there exists $b \in A_P$ such that*

$$(\gamma y_1 x y_2)_{A_P} \in {}^-A_P b x_{A_P} \quad \text{for all } y_1, y_2 \in \Omega, x \in \mathfrak{S}, \gamma \in \Gamma.$$

PROOF. Applying $\alpha'_{P,i}$ to $(\gamma y_1 x y_2)_{A_P}/x_{A_P}$ yields a number which is < 1 times a bounded factor for each i. We just select $b \in A_P$ such that $b^{\alpha'_{P,i}} \geqq c_1$ for all i, with a suitably large constant c_1. Let z be any quotient

$$z = (\gamma y_1 x y_2)_{A_P}/x_{A_P}$$

as in Lemma 6.1. Then zb^{-1} is $\mathcal{S}(\mathfrak{n}_P)'$-log negative, i.e. lies in ^-A_P, as desired.

The above estimates can be made explicit on SL_2 in terms of the matrix coordinates. Cf. the example in Chapter 2 §1.

1.7. Parabolic integral formulas

First we fix some terminology and recall an abstract nonsense formula. Let G be a unimodular locally compact group and K a closed subgroup. *If a left G-invariant measure exists on G/K, we call such a measure a* **Haar measure**. We then have what we shall call the **twisted Fubini formula**

$$\int_G f(x)dx = \int_{G/K} \left[\int_K f(xk)dk \right] dx_{G/K} \quad \text{for } f \in C_c(G). \tag{1}$$

If any two of the measures $dx, dk, dx_{G/K}$ are given, then the third is uniquely determined to make the formula valid without an extra constant factor.

BASIC ASSUMPTIONS. *Let G be a unimodular locally compact group. Let K, P be closed subgroups and K compact. Let $K_P = K \cap P = P_K$. We assume*

$$G = PK,$$

but not necessarily as direct product. We let Δ_P be the modular character on P, that is for $f \in C_c(P)$,

$$\int_P f(pp_1)dp = \Delta_P(p_1) \int_P f(p)dp.$$

We have $\Delta_P(K_P) = 1$ (because K_P is compact and Δ_P is $\mathbf{R}_{>0}$-valued).

In particular, there is a Haar measure on G/K and G/K_P. Cf. [**Lan 99**], Chapter XVI, Theorem 5.1. Similarly, there is a Haar measure on P/K_P. For more general statements (G not unimodular, K not compact) cf. [**Rag 72**].

Let us now take care of the constant factors.

H1. We fix a Haar measure dx on G.

H2. We let K, K_P and K/K_P have measure 1. This uniquely determines the Haar measure $dx_{G/K}$ on G/K to make the twisted Fubini theorem valid for the inclusion $G \supset K$, in other words

$$\int_G f(x)dx = \int_{G/K} \left[\int_K f(xk)dk \right] dx_{G/K} \quad \text{for } f \in C_c(G). \tag{2}$$

H3. We give P/K_P the measure dp corresponding to the Haar measure on G/K under the P-homogeneous space isomorphism

$$P/K_P \to G/K \quad \text{arising from} \quad G = PK.$$

The three conditions **H1, H2, H3** define the **(K, P)-normalization of the Haar measures.**

Let $f \in C_c(G/K)$, so f is right K-invariant. Let f_P be its pull-back to P under the bijection $P/K_P \to G/K$. Then $f_P \in C_c(P/K_P)$ and from (2) we get

$$\int_G f(x)dx = \int_{G/K} f(x_{G/K})dx_{G/K}. \tag{3}$$

We obtain further that

$$\int_G f(x)dx = \int_P f_P(p)dp. \tag{4}$$

PROOF. Starting with the right side of (3), we have

$$\begin{aligned}\int_{G/K} f(x_{G/K})dx_{G/K} &= \int_{PK/K} f(x_{G/K})dx_{G/K} \\ &= \int_{P/K_P} f_{P/K_P}(x_{P/K_P})dx_{P/K_P} \\ &= \int_P f_P(p)dp. \quad \text{qed}\end{aligned}$$

The above abstract nonsense applies to our reduced parabolic $P = U_P A_P G_P$ by Theorem 1.1. We call (u, a, g) with $u \in U_P, a \in A_P$, and $g \in G_P$ the **P-parabolic coordinates** of an element $p \in P$, and also of the element $x \in G/K$ corresponding to p under the parabolic coordinate map. By standard Haar measures computations (e.g. [**JoL 01a**] Chapter I, Propositions 2.1 and 2.2, (see also Chapter V, Lemmas 3.1, 3.2, 3.3) we obtain the **parabolic coordinates integration formula**

$$\boxed{\int_G f(x)dx = \int_{U_P}\int_{A_P}\int_{G_P} f_P(uag)\delta_P(a)^{-1}dudadg}. \tag{5}$$

We have written δ_P instead of Δ_P, but until Chapter 3, we don't need any property of this function except the property that it is the modular character. On the other hand, from Proposition 2.2 mentioned above and the subsequent Proposition 2.4, one sees that if $\mathbf{c}_{\mathfrak{n}_P}(a)$ denote the conjugation action of an element $a \in A_P$ on $\mathfrak{n}_P$, then

$$\Delta_P(a) = |\det \mathbf{c}_{\mathfrak{n}_P}(a)|.$$

Since $\mathfrak{n}_P = \bigoplus \mathfrak{n}_{P,\alpha}$ is the direct sum of A-eigenspaces $\mathfrak{n}_{P,\alpha}$ with eigencharacter α, letting $m(\alpha) = \dim \mathfrak{n}_{P,\alpha}$, the determinant is given by

$$\Delta_P(a) = \delta_P(a) = \prod_{\alpha\in\mathcal{R}(\mathfrak{n}_P)} a^{m(\alpha)\alpha} \quad \text{even for } a \in A.$$

On $\mathrm{SL}_n(\mathbf{C})$, we have $m(\alpha) = 2$. As a matter of notation, we defined independently

$$\rho_P = \frac{1}{2}\sum_{\alpha\in\mathcal{R}(\mathfrak{n}_P)} m(\alpha)\alpha = \sum_{\alpha\in\mathcal{R}(\mathfrak{n}_P)} \alpha \quad \text{on} \quad \mathrm{SL}_n(\mathbf{C}).$$

Thus

$$\delta_P(a) = a^{2\rho_P} = a^{\tau_P}.$$

In the present theory, the critical strip consists of those characters such that $0 < \mathrm{Re}(\zeta) < 2\rho_P$, so ρ_P is at the center of the real part of the critical strip. It's like 1/2 and 1 respectively in the one-variable case.

In the integral formula, the space C_c is the standard space of test functions for measure theoretic considerations. By general measure theory, the above formulas extend to functions in L^1. In practice, we shall apply the formulas to continuous functions which are sufficiently rapidly decreasing at infinity. What "sufficiently rapidly" means will be discussed below.

We give immediately an application of the parabolic coordinates integration formula. For simplicity, we first give the application with respect to an Iwasawa decomposition $G = UAK$. Cf. $\mathrm{SL}_2(\mathbf{R})$, Chapter 4, §3, Theorem 4.

PROPOSITION 7.1. *Let χ be a character on A, extended to G via the Iwasawa product decomposition, that is $\chi(uak) = \chi(a)$. Let $\varphi \in C_c(K\backslash G/K)$ be K-bi-invariant and even, that is $\varphi(z) = \varphi(z^{-1})$ for $z \in G$. Then*

$$\chi * \varphi = (\chi * \varphi)(e)\chi.$$

In other words, χ is an eigenfunction of right convolution by $C_c^\infty(K\backslash G/K)_{\text{even}}$.

REMARK. Since φ is assumed even, it could additionally be assumed only K-invariant on the right, or on the left. The bi-invariance then follows from the even condition. The above reference gives a slightly longer proof without the evenness assumption.

PROOF. We have from the definitions and the hypothesis φ even,

$$\chi * \varphi(x) = \int_G \chi(xy^{-1})\varphi(y)dy = \int_G \chi(y)\varphi(x^{-1}y)dy.$$

Use Iwasawa coordinates $y = uak, dy = \delta^{-1}(a)dudadk$, and let $x = vbk_1$. Then the last integral is

$$= \int_U \int_A \int_K \chi(a)\varphi(k_1^{-1}b^{-1}v^{-1}uak)\delta^{-1}(a)dudadk.$$

By the left invariance of the du integral, we can delete v^{-1} in the integral. We make the translation $a \mapsto ba$, preserving the integral, and use the character property

$$\chi(b) = \chi(b)\chi(a) \quad \text{and} \quad \delta^{-1}(ba) = \delta^{-1}(b)\delta^{-1}(a).$$

Finally, we use the definition of δ as the Haar character on UA, so making the change of variables $b^{-1}ub \mapsto u$ introduces the factor $\delta(b)$ in the integral, canceling $\delta^{-1}(b)$ which appeared in the previous step. What is left is

$$\chi(b)\int_U \int_A \int_K \chi(a)\varphi(uak)\delta^{-1}(a)dudadk = (\chi * \varphi)(e)\chi(b),$$

which proves the proposition.

Note that the above result is a piece of Haar measure abstract nonsense, holding for a Lie group with a weak Iwasawa decomposition as defined in [**JoL 01a**], Chapter

I. We restate the proposition in the context in which we shall use it, for parabolic decompositions.

PROPOSITION 7.2. *Let P be a standard reduced parabolic, and let χ be a character on A_P, extended to G via projection on A_P using parabolic coordinates. Let $\varphi \in C_c(K\backslash G/K)$ be even. Then*

$$\chi * \varphi = (\chi * \varphi)(e)\chi.$$

PROOF. The proof is the same, using the parabolic coordinates integration formula (5) instead of the Iwasawa decomposition.

The above propositions are analogues of the standard eigenfunction properties for spherical functions, cf. [**JoL 01a**], Chapter IV, Theorem 5.1. They are also variations of Selberg's eigenfunction property with point-pair invariants, because the function $(x, y) \mapsto \varphi(y^{-1}x)$ is a point pair invariant, cf. [**JoL 01a**], Chapter IV, Proposition 2.2. The propositions are related to a general theorem of Harish-Chandra [**Har 59**], see also [**Bor97**], Theorem 2.14, which applies to a wider class of functions, but only asserts the existence of one φ for which such functions are eigenfunctions under right convolution with φ. In our present context, we could give a much shorter proof, with a natural space of functions φ. Furthermore, the proof works for a situation which axiomatizes an Iwasawa decomposition as discussed in [**JoL 01a**] Chapter I, §1.

REMARK. The auxiliary function φ can obviously be selected so that

$$(\chi * \varphi)(e) \neq 0,$$

so we recover χ as a convolution with elements of $C_c^\infty(K\backslash G/K)_{\text{even}}$. Indeed, if χ is real, $\chi = \chi_\lambda$ with λ real valued, then $\chi(e) = 1$ and $\chi > 0$, so any function φ as in the proposition, and such that $\varphi \geqq 0, \varphi(e) > 0$, will be such that $\chi * \varphi(e) \neq 0$. This gives some uniformity in the proposition.

We now give a second application of parabolic coordinates. This is a special case of the general result in [**JLS 02**]. Let $\mathbf{S}_G$ be the spherical transform. Harish-Chandra's commutative diagram (see [**JoL 01a**], Chapter 3, Proposition 5.1) tells us that $\mathbf{S}_G = \mathbf{M}_G\mathbf{H}_G$ is the composite of the Harish transform $\mathbf{H}_G$ and the Mellin transform $\mathbf{M}_G$, defined on the appropriate spaces, e.g. the Harish-Chandra Schwartz space HCS or better HCS^1. See [**JoL 01a**], Chapter X, Theorem 5.6 and Chapter XI, Theorem 2.3 (the L^1 extension by Trombi-Varadarajan).

We let $V = UA$ and $V_P = U_P A_P$. Write $\zeta \in \mathfrak{a}_{\mathbf{C}}^\vee$ as $\zeta = \zeta_P + \zeta_{G_P}$ in accordance with Lemma 4.3. If $v = ua$ then $dv = a^{-2\rho}duda$, and similarly with the index P. Then from $\mathbf{S}_G = \mathbf{M}_G\mathbf{H}_G$, we get

$$(6) \qquad (\mathbf{S}_G f)(\zeta) = \int_V f(v)(v_A)^{\zeta+\rho}dv = \int_U\int_A f(ua)a^{\zeta-\rho}duda.$$

On $\mathrm{SL}_n(\mathbf{C})$, by the product structure of G_P, the spherical transform $\mathbf{S}_{G_P}$ is defined by the similar formula for each factor, with variables $v_{G_P} \in V_{G_P}$, $\zeta_{G_P} \in \mathfrak{a}_{G_P,\mathbf{C}}^\vee$, and ρ_{G_P}. It will be convenient to write f_ζ for the function defined by

$$f_\zeta(x) = f(x)x_A^\zeta.$$

We use a similar notation if $\zeta_P \in \mathfrak{a}_{P,\mathbf{C}}^\vee$, taking the projection x_{A_P} on A_P. We then define the **parabolic spherical projection** $\mathbf{S}_{G_P}^G$ by

$$\begin{aligned}(\mathbf{S}_{G_P}^G f_{\zeta_P})(g) &= \int_{V_P} f(v_P g)(v_P)_{A_P}^{\zeta_P+\rho_P}dv_P \\ &= \int_{U_P}\int_{A_P} f(u_P a_P g)a_P^{\zeta_P-\rho_P}du_P da_P. \qquad (7)\end{aligned}$$

Here $g \in G_P/K_{G_P}$ and $\mathbf{S}_{G_P}^G$ maps functions on $K\backslash G/K$ to functions on the space $K_{G_P}\backslash G_P/K_{G_P}$. The K_{G_P}-invariance on the left comes from the commutativity of a_P and k_{G_P}, writing $u_P a_P k_{G_P} = k_{G_P}k_{G_P}^{-1}u_P k_{G_P} a_P$, and then using the fact that du_P is k_{G_P}-conjugation invariant, and f is left K-invariant. We then have the **parabolic spherical decomposition**

$$\mathbf{S}_{G_P}(\mathbf{S}_{G_P}^G f_{\zeta_P})(\zeta_{G_P}) = (\mathbf{S}_G f)(\zeta_P + \zeta_{G_P}) = (\mathbf{S}_G f)(\zeta), \qquad (8)$$

which merely expresses Fubini's theorem (referring to Lemma 4.4)

$$(\mathbf{S}_G f)(\zeta) = \int_{U_{G_P}}\int_{A_{G_P}}\int_{U_P}\int_{A_P} f(u_P u_{G_P} a_P a_{G_P})a_P^{\zeta_P-\rho_P}a_{G_P}^{\zeta_{G_P}-\rho_{G_P}}du_P du_{G_P} da_P da_{G_P}.$$

Here we use that a_P and u_{G_P} commute. The formula is valid under appropriate conditions guaranteeing absolute convergence of the integral.

We may then apply the above formalism for the heat Gaussian, stemming from Gangolli [**Gan 68**]. See [**JoL 01a**] Chapter X, §7, Chapter XII, §5, and also Chapter 2, §2 below and the Appendix. For our purposes here, we define the **heat Gaussian** $\mathbf{g}_{G,t}$ on G/K to be the inverse spherical transform of the Gaussian $E_{G,t}$ on $\mathbf{i}\mathfrak{a}^\vee = \mathbf{i}\mathfrak{a}_G^\vee$ normalized by

$$E_{G,t}(\zeta_G) = \exp((\zeta_G^2 - \rho_G^2)t).$$

Thus

$$\mathbf{S}_G\mathbf{g}_{G,t} = E_{G,t} \quad \text{or also} \quad \mathbf{g}_{G,t} = \mathbf{S}_G^{-1}E_{G,t}.$$

Similarly we define $\mathbf{g}_{G_P,t}$ on G_P/K_{G_P}, so $\mathbf{S}_{G_P}\mathbf{g}_{G_P,t} = E_{G_P,t}$.

THEOREM 7.3. *We have the relation*

$$\mathbf{g}_{G_P,t} = e^{\rho_P^2 t}\mathbf{S}_{G_P}^G \mathbf{g}_{G,t}.$$

PROOF. By (8) and Lemma 5.1, it follows that the two functions on the right and left of the desired equation have the same image $E_{G_P,t}$ under the spherical transform $\mathbf{S}_{G_P}$ on G_P/K_{G_P}. Hence they are equal, as was to be shown.

The heat kernel $\mathbf{K}_{\mathbf{X}_G,t}$ on $\mathbf{X}_G = G/K$ is given by

$$\mathbf{K}_{\mathbf{X}_G,t}(x,y) = \mathbf{g}_{G,t}(y^{-1}x),$$

so Theorem 7.3 yields the analogous formula relating the heat kernel on $\mathbf{X}_{G_P}$ with the heat kernel on $\mathbf{X}_G$.

CHAPTER 2

Eisenstein Series

In the first chapter, we described reduced parabolic subgroups more general than the subgroup UA coming from the Iwasawa decomposition $G = UAK$. These subgroups contain UA, and the world is made up so that they provide a complete system for the continuous part of the Fourier decomposition on $\Gamma\backslash G/K$, where for us $\Gamma = \mathrm{SL}_n(\mathbf{Z}[\mathbf{i}])$, but the situation is typical. In addition, they allow what is called parabolic induction, i.e. an inductive procedure which reduces spectral decomposition to groups of lower dimension, of the same type as G. Thus they contain all the information relevant for us about $\Gamma\backslash G$, in a neat package.

On a compact quotient $\Gamma\backslash G$ with other types of Γ's than we consider, Fourier decomposition looks as it does on the circle. There is some orthonormal basis for whatever space one wants, such that a reasonable function f has the Fourier series

$$f = \sum \langle f, \psi_k \rangle \psi_k,$$

in terms of this basis and Fourier coefficients. In our case, of course, the quotient is not compact, and there has to be added a continuous part, expressed as an integral, or rather a sum of integrals, indexed by the (reduced) parabolics.

The integrand taking the place of the orthonormal basis above turns out to be an Eisenstein series, in which the heat kernel enters in an essential way. It is a gadget which encodes into one object an infinite amount of spectral information. Thus the continuous part of spectral decomposition on $\Gamma\backslash G$ can be expressed as a finite sum, taken over the finite set of parabolic subgroups rather than an infinite sum over an orthonormal system of some sort. Conceptually and technically, the introduction of the heat Eisenstein series thus simplifies and extends the classical theory, where Eisenstein series are built up from characters and automorphic forms.

2.1. The character Eisenstein series

Let P be a reduced standard parabolic of $G = \mathrm{SL}_n(\mathbf{C})$. Let

$$\Gamma = \mathrm{SL}_n(\mathbf{Z}[\mathbf{i}]) \text{ and } \Gamma_P = \Gamma \cap P.$$

Let f be a function on G/K which is left Γ_P-invariant. Its **trace** $\mathrm{Tr}_{\Gamma_P\backslash\Gamma}(f)$ is defined by

$$(\mathrm{Tr}_{\Gamma_P\backslash\Gamma} f)(x) = \sum_{\gamma\in\Gamma_P\backslash\Gamma} f(\gamma x).$$

This trace occurs systematically, in various contexts of the function f. In certain contexts, it is called an **Eisenstein trace** or **Eisenstein series**. We begin with the most classical case involving a character.

Let χ be a character on A_P. For $\zeta = \zeta_P \in \mathfrak{a}^\vee_{P,\mathbf{C}}$, we let

$$\chi = \chi_\zeta \quad \text{so that for } a \in A_P \text{ we have} \quad \chi_\zeta(a) = a^\zeta.$$

For $x \in G$, we recall the parabolic component x_{A_P} coming from Chapter 1, Theorem 4.1. We define the **character Eisenstein series** with a character $\chi = \chi_\zeta$ by

$$\begin{aligned} E_P(\chi_\zeta)(x) = E_P(\chi_\zeta, x) = E(\zeta, x) &= \sum_{\gamma \in \Gamma_P \backslash \Gamma} (\gamma x)^\zeta_{A_P} \\ &= \sum_{\gamma \in \Gamma_P \backslash \Gamma} \chi_\zeta((\gamma x)_{A_P}). \end{aligned} \tag{1}$$

This definition depends on Chapter 1, §4, (2), which tells us that $(\gamma x)_{A_P}$ depends only on the coset $\Gamma_P \gamma$. Thus we are dealing with the function

$$f(x) = (x_{A_P})^\zeta,$$

in other words, we view χ_ζ as being extended to a function on G/K by composition with the projection on A_P. This function is left Γ_P-invariant.

We shall use the real part and positivity of characters discussed in Chapter 1, §5. We had the dual basis $\{\alpha'_{P,1}, \ldots, \alpha'_{P,r_P}\}$ of $\mathfrak{a}^\vee_P$, such that if a character $\zeta \in \mathfrak{a}^\vee_{P,\mathbf{C}}$ is expressed in terms of this basis

$$\zeta = \sum s_i \alpha'_{P,i} \quad s_i \in \mathbf{C},$$

then $\mathrm{Re}(\zeta) > 0$ if and only if $\mathrm{Re}(s_i) > 0$ for all i. By definition,

$$a^{2\rho_P} = \delta_P(a).$$

THEOREM 1.1. *For* $\mathrm{Re}(\zeta) > 2\rho_P$, *the Eisenstein series* $E_P(\chi_\zeta)$ *is absolutely convergent, uniformly for* x *in a compact subset of* G.

This theorem for a much wider class of groups is proved in [**Lgds 76**], see also [**Har 68**]. In this reference, Harish-Chandra makes two translations by ρ_P, and uses the anti-Iwasawa decomposition. This accounts for the differently described domains of convergence in [**Har 68**], Chapter II, Lemma 23 p.26 and Corollary 1 p.28. One of the translations occurs in the definition before §2, p.26, which we found very misleading. The other translation occurs in the definition of φ_λ, p.26 also. Preserving, as we do, the notion that $2\rho_P$ is at the edge of the critical strip emphasizes an essential structure of the situation. Our normalization corresponds to the classical normalization of analytic number theory, namely, the edge of the critical strip for the Riemann zeta function is at $s = 1$.

Harish-Chandra's normalization with the translations with ρ_P stems from an analogy with the theory of spherical functions, where the translations are natural. Although the analogy does exist and deserves to be further investigated, at the moment we are giving priority to the classical right half spaces of convergence for what amounts to generalized Dirichlet series. The forthcoming further development of the theory may shed further light on the respective roles of these normalizations.

Example. $G = \mathrm{SL}_2(\mathbf{R})$. For this example, see [**Lan 75/85**], Chapter XIII, §3, Lemma 2, and [**Bor 97**], Theorem 10.4. In this case, G/K has the model of the upper half plane, with variable $z = x + \mathbf{i}y, y > 0$. There is only one parabolic UA, from the Iwasawa decomposition $G = UAK$. Note that for $b \in \mathbf{R}, a_1 > 0$,

$$\begin{pmatrix} 1 & b \\ 0 & 1 \end{pmatrix} \begin{pmatrix} a_1 & 0 \\ 0 & a_1^{-1} \end{pmatrix} (\mathbf{i}) = a_1^2\mathbf{i} + b.$$

Thus $y = a_1^2 = a^\alpha = a^{2\rho}$ if $\rho = \alpha/2$. Also $x = b$. On the other hand, if

$$\gamma = \begin{pmatrix} * & * \\ c & d \end{pmatrix} \quad \text{then} \quad \mathrm{Im}(\gamma(z)) = \frac{\mathrm{Im}(z)}{|cz+d|^2} = \frac{y}{|cz+d|^2} = (\gamma z)_A^\alpha.$$

Finally Γ_U consists of all the matrices $\begin{pmatrix} 1 & b \\ 0 & 1 \end{pmatrix}$ with $b \in \mathbf{Z}$, so one finds immediately a bijection

$$\Gamma_U \backslash \Gamma \longrightarrow \text{relatively prime pairs } (c, d) \in \mathbf{Z}^2.$$

Take α as the basis for $\mathfrak{a}^\vee$, and write a character as $s\alpha$ with $s \in \mathbf{C}$. Then the Eisenstein series is

$$E(\chi_{s\alpha})(z) = \sum_{(c,d)} \frac{y^s}{|cz+d|^{2s}} = \sum_{(c,d)} \frac{a^{2s\rho}}{|cz+d|^{2s}}.$$

A trivial estimate (of the number of pairs $(m_1, m_2) \in \mathbf{Z}^2$ in an annulus of width 1 and radius $n \to \infty$) shows that the series is absolutely convergent for $\mathbf{R}e(s) > 1$. On the other hand, $\langle \alpha, \alpha \rangle = 2$, so the dual basis of α is $\alpha' = \alpha/2$. Then $s\alpha = s'\alpha/2$ and $s' = 2s$. Thus in terms of s', the half plane of convergence is $\mathbf{R}e(s') > 2$. This is the normalization used in the general case.

The rest of this section is devoted to the proof of Theorem 1.1, which also determines an order of growth of the Eisenstein series. A character ζ can be written as $\zeta = \xi + \mathbf{i}\lambda$ where $\xi, \lambda \in \mathfrak{a}_P^\vee$ are real, $\xi = \mathrm{Re}(\zeta)$, $\lambda = \mathrm{Im}(\zeta)$. Convergence depends only on the real part, so we suppose $\zeta = \xi$ is real. We let χ also denote the extension of χ_ξ to G via the projection on A_P, to simplify the notation, instead of writing χ_G. Thus for $y \in G$,

$$\chi(y) = (y_{A_P})^\xi.$$

By Chapter 1, Proposition 7.2, there exists a semipositive even function $\varphi \in C_c(K\backslash G/K)$ such that $\chi * \varphi = \chi$. We shall prove a more explicit version of Theorem 1.1, giving a dependence on x and ξ.

LEMMA 1.2. *There exists an integer $M' = M'(G, \Gamma)$ such that for all even semipositive $\varphi \in C_c(K\backslash G/K)$ with support in a compact set Ω, we have for all $x \in G$,*

$$\sum_{\gamma \in \Gamma_P \backslash \Gamma} (\chi * \varphi)(\gamma x) \leqq c_1(\varphi) \|x\|^{M'} \int_{\Gamma_P \backslash \Gamma x \Omega} \chi(y) dy.$$

The constant $c_1(\varphi)$ is bounded by the sup norm $\|\varphi\|_\infty$ times a constant $c_1'(\Omega)$.

PROOF. By the evenness of φ, the sum in the lemma is equal to

$$\begin{aligned}\sum_{\gamma\in\Gamma_P\backslash\Gamma}\int_G \chi(\gamma xy)\varphi(y)dy &= \sum_{\gamma\in\Gamma_P\backslash\Gamma}\int_G \chi(y)\varphi(x^{-1}\gamma^{-1}y)dy\\ &= \sum_{\gamma\in\Gamma_P\backslash\Gamma}\int_{\Gamma_P\backslash G}\sum_{\eta\in\Gamma_P}\chi(y)\varphi(x^{-1}\gamma^{-1}\eta^{-1}y)dy\\ &= \int_{\Gamma_P\backslash G}\sum_{\gamma\in\Gamma}\chi(y)\varphi(x^{-1}\gamma^{-1}y)dy\\ &\leqq c_1(\varphi)\|x\|^{M'}\int_{\Gamma_P\backslash\Gamma x\Omega}\chi(y)dy,\end{aligned}$$

by using Chapter 1, Lemma 2.4, and noting that $x^{-1}\gamma^{-1}y\in\Omega$ implies $y\in\Gamma x\Omega$. This proves the lemma.

LEMMA 1.3. *Let $\mathfrak{S}$ be a Siegel set in G and Ω a compact set. There exists $b\in A_P$ such that for $\xi\in\mathfrak{a}_P^\vee$, $\xi>2\rho_P$, and all $x\in\mathfrak{S}$, with $\chi(y)=(y_{A_P})^\xi$,*

$$\int_{\Gamma_P\backslash\Gamma x\Omega}\chi(y)dy\leqq \mathrm{vol}(\Gamma_P\backslash U_PG_P)x_{A_P}^{\xi-2\rho_P}b^{\xi-2\rho_P}c_2(\xi),$$

where $c_2(\xi)$ is given by

$$c_2(\xi)=\prod_{\alpha\in\mathcal{S}(\mathfrak{n}_P)}\langle\xi-2\rho_P,\alpha\rangle^{-1}.$$

REMARK. The constant $c_2(\xi)$ blows up as $\xi\to 2\rho_P$, as it should. In classical language, rm $\mathrm{Re}(\zeta)=2\rho_P$ is the edge of the critical strip.

PROOF. By Lemma 6.2 of Chapter 1, there exists $b\in A_P$ such that

$$(\Gamma xy_2)_{A_P}\in {}^-A_Pbx_{A_P}\quad\text{for all}\quad y_2\in\Omega K, x\in\mathfrak{S},\gamma\in\Gamma.$$

We estimate the integral in parabolic coordinates. Since $\chi(y)$ depends only on the A_P-component y_{A_P}, the domain of integration can be replaced by the larger domain

$$\Gamma_P\backslash U_PG_P\,{}^-A_Pbx_{A_P}K.$$

Since $\Gamma_P=\Gamma_{U_P}\Gamma_{G_P}$, and since $\Gamma_{U_P}\backslash U_P$ is compact, while $\Gamma_{G_P}\backslash G_P$ has finite measure, the partial integral

$$\iint_{\Gamma_P\backslash U_PG_P}du_Pdg_P$$

is finite, actually equal to the product of the volume of the two quotient spaces, but the precise evaluation is irrelevant here. It follows that the desired integral up to that constant factor is estimated by the A_P-component integral

$$\int_{{}^-A_Pbx_{A_P}}(abx_{A_P})^\xi(abx_{A_P})^{-2\rho_P}da.$$

The second factor in the integrand is the Jacobian factor going from Haar measure to the parabolic coordinates measure. The integrand is a product of characters, which are homomorphisms, so the factor $(bx_{A_P})^{\xi-2\rho_P}$ comes out of the integral. The remaining integral is

$$\int_{{}^-A_P} a^{\xi-2\rho_P}\,da = c_2(\xi), \tag{2}$$

which we now evaluate to find the value as stated in the lemma.

Let $\{\alpha_{P,1},\dots,\alpha_{P,r_P}\}$ be the elements of $\mathcal{S}(\mathfrak{n}_P)$, and let $\{\alpha'_{P,1},\dots,\alpha'_{P,r_P}\}$ be the dual basis of $\mathfrak{a}_P^\vee$. We have

$$\xi - 2\rho_P = \sum s_{P,i}\alpha'_{P,i} \quad \text{with} \quad s_{P,i} = \langle \xi - 2\rho_P, \alpha_{P,i}\rangle.$$

Hence

$$\int_{{}^-A_P} a^{\xi-2\rho_P}\,da = \int_{{}^-A_P} \prod_i e^{s_{P,i}\alpha'_{P,i}(\log a)}\,da$$
$$= \prod_i \int_{-\infty}^{0} e^{s_{P,i}\alpha'_{P,i}}\,d\alpha'_{P,i}$$

which is trivially evaluated to give the stated value. This concludes the proof.

Putting the preceding lemmas together, we obtain a uniform estimate.

THEOREM 1.4. *For* $\mathrm{Re}(\zeta) > 2\rho_P$ *the Eisenstein series* (1) *has hermitian polynomial growth (i.e. Lie exponential growth). More precisely, given a Siegel set* $\mathfrak{S}$ *and a compact set* Ω *in* G, *there exists an integer* M' *and a locally bounded function* c *on the half space* $\mathrm{Re}(\zeta) > 2\rho_P$ ($\zeta \in \mathfrak{a}_{P,\mathbf{C}}^\vee$) *such that for* $\xi = \mathrm{Re}(\zeta) > 2\rho_P$,

$$\sum_{\gamma\in\Gamma_P\backslash\Gamma} |(\gamma y_1 x)_{A_P}^{\zeta}| \leqq c(\xi)\|x\|^{M'} x_{A_P}^{\xi-2\rho_P}$$

for all $x \in \mathfrak{S}$ *and* $y_1 \in \Omega$.

The general case is treated in [**Har 68**], Lemma 24 on p.20. On $\mathrm{SL}_2(\mathbf{R})$, see [**Bor 97**], Theorem 10.4.

COROLLARY 1.5. *The Eisenstein series* $E(\zeta, x)$ *is complex analytic in* ζ *for* $\mathrm{Re}(\zeta) > 2\rho_P$, *and* C^∞ *in* $x \in G$.

PROOF. The uniformities of Theorem 1.4 show that one can differentiate the series term by term in the given region.

The need for other Eisenstein series

The definition of an Eisenstein series in general is just the $\Gamma_P\backslash\Gamma$-trace of a left Γ_P invariant function f, namely

$$E_P(f)(x) = \sum_{\Gamma_P\backslash\Gamma} f(\gamma x).$$

Taking a character for f is the most classical way, but as we have just seen, as a function of the character, the series converges only in a half space. One can then do two things which help with the convergence and contribute additional structure. One can replace the character by a suitably rapidly decreasing function, or one can twist the character Eisenstein series by another function which is suitably rapidly decreasing. The second alternative is the deeper one, but we shall also use the first alternative, the other functions playing the role of "test functions", ubiquitous in analysis. The most common space of test functions is the space of continuous (or C^∞) functions with compact support. For our purposes, we need a less restrictive space, and we found the Gauss space to serve our needs. We define it in §3 and apply it in the subsequent sections. Ultimately, we want a result which involves only the character Eisensteins twisted by the heat kernel, as in Theorem 5.5, which is the first main thing we are after for this chapter.

The next thing we are after will arise when we use a specific uniqueness property of solutions of the heat equation, and culminates with Theorems 5.1 and 5.2 of Chapter 4, leading into the analytic continuation of the heat and character Eisenstein series.

2.2. Twists of character Eisenstein series

The character Eisenstein series are generalized Dirichlet series, similar to sums $\sum a_n\lambda_n^{-s}$. However, instead of summing over positive integers n, one sums over elements of a discrete group Γ or cosets $\Gamma_P\backslash\Gamma$. Furthermore, instead of one variable s, one has in effect several complex variables when ζ is expressed linearly in terms of a basis. In the theory of Dirichlet series, if $\{b_n\}$ is a sequence, then the series $\sum b_n a_n\lambda_n^{-s}$ is called a **twist** of the Dirichlet series. We shall extend the notion of such twists to Eisenstein series.

We shall also use another structure, the product structure on G. Quite generally, suppose a space X is expressed as a product

$$X = X_1 \times X_2.$$

Let f_1 be a function on X_1 and f_2 a function on X_2. We define the **tensor product function** $f_1 \otimes f_2$ on X by

$$(f_1 \otimes f_2)(x_1, x_2) = f_1(x_1)f_2(x_2).$$

One may view f_1 as extended to X by composing with projection on X_1, and similarly for f_2, so one may write occasionally f_1f_2 instead of writing explicitly the tensor product sign. A fully correct notation would be

$$(f_1 \circ \mathrm{pr}_1)(f_2 \circ \mathrm{pr}_2)$$

where pr_i is the projection on X_i.

Under the P-homogeneous space isomorphism

$$P/K_{G_P} \xrightarrow{\approx} G/K,$$

G/K has the product structure

$$U_P \times A_P \times G_P/K_{G_P} \xrightarrow{\approx} G/K.$$

The parabolic coordinates integration formula also tells us that for the homogeneous space (Haar) measure on G, the map is an isomorphism for the measure on the product $\delta_P^{-1}(a)dudadg$.

The function f used to make up an Eisenstein series, i.e., $\mathrm{Tr}_{\Gamma_{G_P}\backslash\Gamma}(f)$, will be built up partly from tensor products of functions on the three different factors. A common pattern will be the following. We take F on G_P/K_{G_P}, φ on A_P, and ψ on $\Gamma\backslash G/K$, so ψ is on G with left Γ-invariance and right K-invariance. Let $f = (F \otimes \varphi)\psi$. Then readers should keep in mind that

$$\mathrm{Tr}_{\Gamma_{G_P}}((F \otimes \varphi)\psi) = (\mathrm{Tr}_{\Gamma_{G_P}}(F) \otimes \varphi)\psi.$$

Thus φ, ψ act like "constants" with respect to the Γ_{G_P}-trace.

We start with the most important special case when the Eisenstein series will be taken with a tensor product of a function on G_P/K_{G_P} and a character on A_P.

One variable twists

As remarked in Chapter 1, §4, dealing with reduced parabolics yields

$$K_P = K_{G_P}.$$

We let $\mathbf{X}_{G_P} = G_P/K_{G_P}$. We shall consider functions F on $\mathbf{X}_{G_P}$ and also functions on $\mathbf{X}_{G_P}$ which are left invariant by Γ_{G_P}. We start with the second case, because the first case will be reduced to the second one.

Let F_0 be a function on $\Gamma_{G_P}\backslash G_P/K_P$. Let $\chi = \chi_\zeta$ with $\mathrm{Re}(\zeta) > 2\rho_P$. Then we define the F_0-**twisted Eisenstein** series $E_P(F_0, \chi)$ by the series

$$E_P(F_0, \chi)(x) = \sum_{\Gamma \in \Gamma_P\backslash\Gamma} F_0((\Gamma x)_{\mathbf{X}_{G_P}})\chi((\Gamma x)_{A_P}).$$

We also call this the F_0-**twist** of $E_P(\chi)$. We make the assumption:

The series converges absolutely, uniformly for x in compact sets.

In practice, we shall in fact deal with F_0 bounded, in which case this condition is trivially satisfied, reducing the convergence to that of the character Eisenstein series itself. Note that Γ should also be in the notation, so we should write $E_{P,\Gamma}$ instead of E_P, but Γ will be fixed so we sometimes omit it.

We also note that an Eisenstein series can be viewed as a generalized Dirichlet series.

In practice, we do not start from a Γ_{G_P}-invariant function but start with a function F on $\mathbf{X}_{G_P} = G_P/K_{G_P}$. We obtain a Γ_{G_P}-invariant function by taking the Γ_{G_P}-trace. More precisely, let $F \in C(\mathbf{X}_{G_P}) = C(G_P/K_{G_P})$ be a continuous function. We say that F is an **admissible Eisenstein twister** if the following conditions are satisfied.

ET 1P. The series

$$\mathrm{Tr}_{\Gamma_{G_P}}(F)(g) = \sum_{\eta \in \Gamma_{G_P}} F(\eta g)$$

is absolutely convergent, uniformly on compact sets.

ET 2P. For all $\mathrm{Re}(\zeta) > 2\rho_P$ the double series

$$\sum_{\gamma \in \gamma_P \backslash \gamma} \chi_\zeta((\gamma x)_{A_P}) \sum_{\eta \in \gamma_{G_P}} F((\eta\gamma x)\mathbf{x}_{G_P}),$$

is absolutely convergent, uniformly for x is compact sets.

We note that for $\xi = \mathrm{Re}(\zeta)$,

$$|\chi_\zeta(a)| = \chi_\xi(a),$$

so for each ζ, the absolute value $|\chi_\zeta|$ depends only on the real part of ζ, and the estimate is uniform on each imaginary axis $\xi + \mathbf{i}\mathfrak{a}_P^\vee$.

We keep in mind that $\Gamma_{U_P}\backslash\Gamma_P$ can be identified with Γ_{G_P}. Furthermore, for $\eta \in G_P$ and $x \in G$ we have $(\eta x)\mathbf{x}_{G_P} = \eta(x\mathbf{x}_{G_P})$. Then for $\chi = \chi_\zeta$ and F satisfying **ET 1P** and **ET 2P**, we define the (F, χ)-**Eisenstein series** $E_{P,\gamma,F}(\chi)$ to be the series

$$\begin{aligned}(1) \qquad E_P(\mathrm{Tr}_{\Gamma_{G_P}}(F), \chi)(x) & \sum_{\gamma \in \Gamma_P\backslash\Gamma} \mathrm{Tr}_{\Gamma_{G_P}}(F)((\gamma x)\mathbf{x}_{G_P})\chi((\gamma x)_{A_P}) \\ &= E_{P,\Gamma,F}(\chi)(x) = \sum_{\gamma \in \Gamma_P\backslash\Gamma} \chi((\gamma x)_{A_P}) \sum_{\eta \in \Gamma_{U_P}\backslash\Gamma_P} F((\eta\gamma x)\mathbf{x}_{G_P}).\end{aligned}$$

We also call this the F-**twist** of $E_P(\chi)$. It is often convenient to abbreviate and let

$$F_0 = \mathrm{Tr}_{\Gamma_{G_P}}(F).$$

We shall need to estimate the Eisenstein series, and we note that if $\xi = \mathrm{Re}(\zeta)$,

$$\begin{aligned}E_P(\mathrm{Tr}_{\Gamma_{G_P}}|F|, |\chi_\zeta|) &= \sum_{\gamma \in \Gamma_P\backslash\Gamma} \chi_\xi((\gamma x)_{A_P}) \sum_{\eta \in \Gamma_{U_P}\backslash\Gamma_P} |F((\eta\gamma x)\mathbf{x}_{G_P})| \\ &= E_P(|F_0|, \chi_\xi).\end{aligned}$$

PROPOSITION 2.1. *Suppose that* $\mathrm{Tr}_{\gamma_{G_P}}(|F|)$ *is bounded. Then* **ET 2P** *is satisfied for* $\mathrm{Re}(\zeta) > 2\rho_P$.

PROOF. Immediate from Theorem 1.1.

More generally, we shall also consider (F, φ)-**Eisenstein series**

$$E_P(\mathrm{Tr}_{\Gamma_{G_P}}(F, \varphi)) = \sum_{\gamma \in \Gamma_P \backslash \Gamma} \mathrm{Tr}_{\Gamma_{G_P}}(F)((\gamma x)_{\mathbf{x}_{G_P}})\varphi((\gamma x)_{A_P})$$
$$= \sum_{\gamma \in \Gamma_P \backslash \Gamma} \varphi((\gamma x)_{A_P}) \sum_{\eta \in \Gamma_{U_P} \backslash \Gamma_P} F((\eta\gamma x)_{\mathbf{x}_{G_P}}),$$

with other functions φ on A_P rather than characters. These will be specified as we go along. Mostly the function φ will be taken in the Gauss space, which is going to be defined in the next section.

Example. The Gaussian on G/K. We define the **Gangolli Gaussian** on G/K to be the K-bi-invariant function on G given on A^+ by the **Gangolli formula**:

$$\mathbf{g}_t(a) = \frac{1}{(4\pi t)^{\dim(G/K)/2}} e^{-|\log a|^2/4t} e^{-\rho^2 t} \mathbf{j}(a)^{-1},$$

where $\rho^2 = \langle \rho, \rho \rangle$, *and* $\mathbf{j}$ *is the function given on* $a \in A^+$ *by*

$$\mathbf{j}(a) = \prod_{\alpha \in \mathcal{R}(\mathfrak{n})} \mathbf{j}_\alpha(a) \quad \text{and} \quad \mathbf{j}_\alpha(a) = C_\alpha \frac{\sinh(\alpha(\log a))}{\alpha(\log a)}$$

with the constant

$$C_\alpha = \frac{\langle \alpha, \rho \rangle}{\pi} = \frac{\langle \alpha, \tau \rangle}{2\pi}$$

where $\rho = \frac{1}{2}\tau$, and τ is the trace of the $(\mathfrak{a}, \mathfrak{n})$ representation. In the present case $G = \mathrm{SL}_n(\mathbf{C})$, we have of course $m(\alpha) = 2$, and

$$\dim\ G/K = n^2 - 1.$$

Cf. [**Gan 68**], and [**JoL 01a**], Chapter XII, Theorem 5.1. Having given the values of $\mathbf{g}_t$ on A^+ determines $\mathbf{g}_t$ on all of A (because A^+ is a fundamental domain for the action of the Weyl group W on the set of regular elements in A). Then $\mathbf{g}_t$ is also determined on G by the K-bi-invariance and the polar decomposition $G = KAK$. The above definition is completely elementary. A number of properties can be verified directly, for instance $\mathbf{g}_t$ is even, $\mathbf{g}_t(x^{-1}) = \mathbf{g}_t(x)$ for all $x \in G$.

Note that we may write the expression for $\mathbf{g}_t$ directly on G by using the function σ defined in Chapter 1, §1, namely for $x \in G$, and $N = \dim\ G/K$,

$$\mathbf{g}_t(x) = \frac{1}{(4\pi t)^{N/2}} e^{-\sigma^2(x)/4t} e^{-\rho^2 t} \mathbf{j}(x)^{-1}.$$

Recall that if $x = k_1 b k_2$ is the polar decomposition of x, then $\sigma(x) = |\log\ b|$. The function $\mathbf{g}_t$ gives rise to a function of two variables by letting

$$\mathbf{K}_t(x, y) = \mathbf{g}_t(y^{-1}x).$$

The values of $\mathbf{K}_t$ depend only on $x, y \in G/K$, and $\mathbf{K}_t$ is symmetric, that is

$$\mathbf{K}_t(x, y) = \mathbf{K}_t(y, x).$$

With the theory of spherical functions on G/K, one sees that $\mathbf{K}_t(x, y)$ is the **heat kernel** when G is a complex group [**Gan 68**], in our case $\mathrm{SL}_n(\mathbf{C})$. For a real group, the heat kernel is the spherical inverse transform of the gaussian on euclidean space (cf. also [**JoL 01a**]), but does not split into a formula such as the above. Working with the explicit function $\mathbf{g}_t$ directly makes it unnecessary to know anything else about the heat kernel for the time being.

Note further that the polar Jacobian J in the present complex case is the square of the real polar Jacobian given by

$$J_0(a) = \prod_{\alpha} (a^{\alpha} - a^{-\alpha}).$$

Thus the numerator of $\mathbf{j}$ (up to the constant factor) is the same as the *real* polar Jacobian, so *the numerator of* $\mathbf{j}$ *is the square root of the complex group polar Jacobian.*

Before passing to estimates, we register explicitly the fact that $\mathbf{j}, \mathbf{j}^{-1}$ *are continuously defined for all* $a \in A$, because $(\sinh z)/z$ is continuous on $\mathbf{R}$ (even on $\mathbf{C}$), with limit 1 as z tends to 0.

Estimates are easy on the function $\mathbf{g}_t$, expressed by the formula. The numerator of $\supset$ has Lie exponential linear growth (cf. the terminology of Chapter 1, §1), and its denominator is a Lie polynomial. *Thus* $\mathbf{j}^{-1}$ *has exponential linear decay.* Conversely:

PROPOSITION 2.2. *Let* $c' > 1$. *Then for* $b \in A$, *we have*

$$e^{-\sigma^2(b)}\mathbf{j}^{-1}(b) \gg e^{-c'\sigma^2(b)}.$$

PROOF. This is immediate, because the exponential term grows exponentially quadratic in $|\log\, b|$, but the $\mathbf{j}^{-1}(b)$ decays exponentially linearly in $|\log\, b|$. Note that for $|\log\, b|$ near 0, the $\mathbf{j}^{-1}(b)$ term is bounded.

PROPOSITION 2.3. *Let* F *be* K*-bi-invariant on* G, *and have exponential quadratic decay, i.e. there exists* $c > 0$ *such that*

$$|F(b)| \leqq e^{-c\sigma^2(b)} \quad \textit{for } \sigma(b) \to \infty.$$

Then the trace $\mathrm{Tr}_{\gamma}(F)$ *given by*

$$\mathrm{Tr}_{\Gamma}(F)(y) = \sum_{\gamma \in \Gamma} F(\gamma y)$$

is bounded as a function on G, *or* $\Gamma\backslash G$.

PROOF. This is a special case of Chapter 1, Lemma 2.4.

This last proposition gives us an example of functions whose trace is bounded, and therefore usable for Eisenstein twisting, especially the function $\mathbf{g}_t$ for a fixed value of t.

In §4, we shall return to a more systematic study of Gaussian functions, including the formulation of Fourier inversion in the context of interest to us.

2.3. Two character Eisenstein series

We start with some remarks going from functions of one variable to functions of two variables. Let φ be a K-bi-invariant function on G. We define the **associated point pair invariant** F on $G \times G$ by letting

$$F(x, y) = \varphi(x^{-1}y).$$

Then $F(zx, zy) = F(x, y)$ for all $z \in G$. Furthermore, for $k \in K$,

$$F(xk, y) = F(x, yk) = F(x, y),$$

that is, F is really defined on $G/K \times G/K$. In practice, it will also be true that φ is **even**, that is $\varphi(x^{-1}) = \varphi(x)$ for all $x \in G$, so F is symmetric, namely $F(x, y) = F(y, x)$ for all $x, y \in G$.

PROPOSITION 3.1. *Let φ be an even K-bi-invariant function on G and put*

$$F(x, y) = \varphi(x^{-1}y).$$

Suppose φ has quadratic exponential decay. Define $\mathrm{Tr}_\Gamma(F)$ *by*

$$(\mathrm{Tr}_\Gamma(F))(x, y) = \sum_{\gamma \in \Gamma} F(\gamma x, y) = \sum_{\gamma \in \Gamma} F(x, \gamma y).$$

Then $(x, y) \mapsto \mathrm{Tr}_\Gamma(F)(x, y)$ is bounded on every set $\Omega \times G$ or $G \times \Omega$ with compact Ω.

PROOF. Special case of Chapter 1, Lemma 2.4.

We intend to twist the Eisenstein series by a function of two variables.

Example continued. We shall apply the proposition to the case $\varphi = \mathbf{g}_t$ and the **heat kernel** is $\mathbf{K}_t(x, y) = \mathbf{g}_t(y^{-1}x)$. In this case, the convergence of the trace series is uniform for $t \geqq t_0 > 0$. Furthermore, Propositions 2.1-2.3 and 3.1 apply to G_P instead of G because G_P is merely a product of groups of the same type as G. The group G_P is a product of groups $G_{n_i} (i = 1, \ldots, r+1)$ corresponding to the block decomposition, and so we get the product decomposition

$$\mathbf{K}_{G_P/K_{G_P}} = \bigotimes \mathbf{K}_{n_i}.$$

As with G, we view $\mathbf{K}_{G_P/K_{G_P}}$ as a function on $\mathbf{R}_{>0} \times G_P \times G_P$, which is right $\mathbf{K}_{G_P}$-invariant in the two G_P-variables. As before we let

$$\mathbf{X}_{G_P} = G_P/K_{G_P} \quad \text{and} \quad \mathbf{K}_{\mathbf{X}_{G_P}} = \mathbf{K}_{G_P/K_{G_P}}.$$

For $x, y \in \mathbf{X}_{G_P}$, we have the symmetric expression

$$(1) \quad \mathrm{Tr}_{\Gamma_{G_P}}(\mathbf{K}_{\mathbf{X}_{G_P}})(t, x, y) = \sum_{\eta \in \Gamma_{G_P}} \mathbf{K}_{\mathbf{X}_{G_P}}(t, \eta x, y) = \sum_{\eta \in \Gamma_{G_P}} \mathbf{K}_{\mathbf{X}_{G_P}}(t, x, \eta y).$$

However, there are other interesting possibilities for twists besides the heat kernel, and so we carry out more of the discussion in the more general setting, as in Proposition 2.4.

Let φ_{G_P} be a K_{G_P}-bi-invariant even function on G_P. We assume that φ_{G_P} has quadratic exponential decay on G_P. We let

$$F_{\mathbf{X}_{G_P}}(x,y) = \varphi_{G_P}(y^{-1}x) \text{ on } G_P/K_{G_P}.$$

We can then lift φ_{G_P} and $F_{\mathbf{X}_{G_P}}$ to G resp. $\mathbf{X}$ via the projections on $\mathbf{X}_{G_P}$, arising from the parabolic coordinates product decomposition, Theorem 4.1 of Chapter 1. We define the **two-character twisted Eisenstein series** $E_{P,\Gamma,F}(\chi_1,\chi_2,x,y)$ with characters $\chi_1 = \chi_{\zeta_1}$ and $\chi_2 = \chi_{\zeta_2}$, by

$$E_P(\mathrm{Tr}_{\Gamma_{G_P}}(F_{\mathbf{X}_{G_P}}),\chi_1,\chi_2,x,y) \quad \text{or } E_P(\mathrm{Tr}_{\Gamma_{G_P}}(F_{\mathbf{X}_{G_P}}),\zeta_1,\zeta_2,x,y)$$

$$= \sum_{\gamma_1,\gamma_2\in\Gamma_P\backslash\Gamma} \mathrm{Tr}_{\Gamma_{G_P}}(F_{\mathbf{X}_{G_P}})((\gamma_1 x)_{\mathbf{X}_{G_P}}),(\gamma_2 y)_{\mathbf{X}_{G_P}})\chi_1((\gamma_1 x)_{A_P})\chi_2((\gamma_2 y)_{A_P}),$$

which we can also write in the form

$$= \sum_{\gamma_1,\gamma_2\in\Gamma_P\backslash\Gamma} \mathrm{Tr}_{\Gamma_{G_P}}(F_{\mathbf{X}_{G_P}})((\gamma_1 x)_{\mathbf{X}_{G_P}},(\gamma_2 y)_{\mathbf{X}_{G_P}})(\gamma_1 x)_{A_P}^{\zeta_1}(\gamma_2 y)_{A_P}^{\zeta_2}.$$

This double twist can be written as the iteration of single twists, namely

$$E_P^{(2)}(\mathrm{Tr}_{\Gamma_{G_P}}(F_{\mathbf{X}_{G_P}}),\chi_\zeta)(x,y) = \sum_{\gamma\in\Gamma_P\backslash\Gamma} \mathrm{Tr}_{\Gamma_{G_P}}(F_{\mathbf{X}_{G_P}})(x_{\mathbf{X}_{G_P}},(\gamma y)_{\mathbf{X}_{G_P}})(\gamma y)_{A_P}^{\zeta}.$$

Thus $E_P^{(2)}(\mathrm{Tr}_{\Gamma_{G_P}} F_{\mathbf{X}_{G_P}}),\chi_\zeta)$ is just the $F_{P,x}$-twisted Eisenstein series, with the function

$$F_{P,x}(y) = F_{\mathbf{X}_{G_P}}(x_{\mathbf{X}_{G_P}}, y_{\mathbf{X}_{G_P}}).$$

Then using $E_P^{(1)}$ instead of $E_P^{(2)}$ with respect to the first variable, we have

$$(3) \qquad E_P(\mathrm{Tr}_{\Gamma_{G_P}}(F_{\mathbf{X}_{G_P}}),\chi_{\zeta_1},\chi_{\zeta_2}) = E_P^{(1)}(E_P^{(2)}(\mathrm{Tr}_{\Gamma_{G_P}}(F_{\mathbf{X}_{G_P}}),\chi_{\zeta_1},\chi_{\zeta_2}).$$

The original definition is symmetric in the two variables, so the expression in (2) is symmetric in χ_1,χ_2 and the two implied variables.

The most important case for us here is of course that of the heat kernel, for which we use the notation

$$E_{P,\Gamma,\mathbf{K}}(t,\zeta_1,\zeta_2,x,y)$$

$$= \sum_{\gamma_1,\gamma_2\in\Gamma_P\backslash\Gamma} \mathrm{Tr}_{G_P}(\mathbf{K}_{\mathbf{X}_{G_P},t}((\gamma_1 x)_{\mathbf{X}_{G_P}}(\gamma_1 x)_{A_P}^{\zeta_1}(\gamma_2 y)_{A_P}^{\zeta_2}.$$

PROPOSITION 3.2. *Let φ_{G_P} be as above. Then the double series (2) over $\gamma_1,\gamma_2 \in \Gamma_P\backslash\Gamma$ for the two-character Eisenstein series converges absolutely for* $\mathrm{Re}(\zeta_1), \mathrm{Re}(\zeta_2) > 2\rho_P$. *Replacing $F_{\mathbf{X}_{G_P}}$ by its absolute value and χ_ζ by its absolute value χ_ξ with $\xi = \mathrm{Re}(\zeta)$, given a compact set $\Omega_{\mathbf{X}_{G_P}}$ in $\mathbf{X}_{G_P}$, there is a constant $C(\Omega_{\mathbf{X}_{G_P}},\varphi)$ such that the double series is dominated by this constant times $E_P(\xi_1,x)E_P(\xi_2,y)$, in other words,*

$$E_P(\mathrm{Tr}_{\Gamma_{G_P}}|F_{\mathbf{X}_{G_P}}|,\xi_1,\xi_2,x,y) \leqq C(\Omega_{\mathbf{X}_{G_P}},\varphi)E_P(\xi_1,x)E_P(\xi_2,y)$$

for $x \in G$ such that $x_{\mathbf{X}_{G_P}} \in \Omega_{\mathbf{X}_{G_P}}$ and all $y \in G$.

PROOF. By Proposition 3.1, $\mathrm{Tr}_{\Gamma_{G_P}}(F_{\mathbf{X}_{G_P}})$ is bounded on $\Omega_{\mathbf{X}_{G_P}} \times \mathbf{X}_{G_P}$ by some constant $C = C(\Omega_{\mathbf{X}_{G_P}}, \varphi)$. Then

$$E_P(\mathrm{Tr}_{\Gamma_{G_P}}|F_{\mathbf{X}_{G_P}}|, \xi_2)(x,y) \leqq CE_P(\xi_2, y) \quad \text{for } x_{\mathbf{X}_{G_P}} \in \Omega_{\mathbf{X}_{G_P}} \text{ and } y \in G.$$

We can now apply $E_P^{(1)}$ to conclude the proof.

We see that Proposition 3.2 shifts the question of convergence for the two-variables twisted Eisenstein series to that of the untwisted one-character Eisenstein series.

For a single character χ, we may then specialize the Eisenstein series by putting $\chi_1 = \chi$ and $\chi_2 = \bar{\chi}$ both having real part $> 2\rho_P$. We then obtain an Eisenstein series depending real analytically on χ, namely:

$$(4) \qquad E_P(\mathrm{Tr}_{\Gamma_{G_P}}(F_{\mathbf{X}_{G_P}}), \chi, \bar{\chi})(x,y)$$

$$= \sum_{\gamma_1,\gamma_2 \in \Gamma_P\backslash\Gamma} \mathrm{Tr}_{\Gamma_{G_P}}(F_{\mathbf{X}_{G_P}})((\gamma_1 x)_{\mathbf{X}_{G_P}}, (\gamma_2 y)_{\mathbf{X}_{G_P}})\chi((\gamma_1 x)_{A_P})\bar{\chi}((\gamma_2 y)_{A_P}).$$

If $F_{\mathbf{X}_{G_P}} = \mathbf{K}_{\mathbf{X}_{G_P}}$ is the heat kernel, then of course we call this the P-**heat Eisenstein series.**

REMARK 1. If we express this Eisenstein series as a double sum, summing first over γ_1, and inside this sum take the sum over γ_2, then the sum over γ_2 does not commute with the sum over $\eta \in \gamma_{G_P}$, and hence the P-heat Eisenstein series in one character cannot be written in a straightforward way as an iteration of the one-variable formalism, even though the two character series could so be written.

REMARK 2. Let $G = \mathrm{SL}_2(\mathbf{R})$. Then there is only one parabolic, and $G_P = \{1\}$. Hence one does NOT see the heat kernel come into play at this level, and the heat Eisenstein series is just the product:

$$E_P(1, \chi, \bar{\chi}) = E_P(\chi)E_P(\bar{\chi}).$$

2.4. The Gauss space

Let $\mathfrak{a}$ be a real finite dimensional vector space of dimension r. In the application, $\mathfrak{a} = \mathfrak{a}_P$. The **Schwartz space** of $\mathfrak{a}$ is the space of C^∞ functions f such that f has superpolynomial decay, and so does Df for every invariant partial differential operator D applied to f. We denote this space by $\mathrm{Sch}(\mathfrak{a})$. It is self dual under the Fourier transform, and is closed under convolution product and under the ordinary product of functions. The Fourier transform is bicontinuous. We shall need a subspace of the Schwartz space. *We suppose given a positive definite scalar product on* $\mathfrak{a}$, making $\mathfrak{a}$ into a euclidean space, and we denote $H^2 = \langle H, H\rangle$. This scalar product extends to a $\mathbf{C}$-bilinear product on the complexification $\mathfrak{a}_{\mathbf{C}}$. We define the

Gauss space Gauss($\mathfrak{a}$) to be the vector space of linear combinations of functions of the form $e^{q(H)}$ where

$$q(H) = -cH^2 + \eta(H) + \text{constant},$$

with $c > 0$ and some linear function $\eta \in \mathfrak{a}^\vee_{\mathbf{C}}$. Thus q is an arbitrary quadratic polynomial with negative homogeneous term of degree 2. We call such q **Gaussian polynomials.**

The Gauss space is an algebra (closed under products), invariant under translations, and contained in the Schwartz space.

In other words,

$$H \mapsto e^{-cH^2} e^{\eta(H)}$$

is in the Schwartz space. It is the restriction to the "real axis" $\mathfrak{a}$ of the space of entire functions $e^{q(Z)}$ where $q(Z)$ is the quadratic function on $\mathfrak{a}^\vee_{\mathbf{C}}$ whose restriction to $\mathfrak{a}$ is a polynomial $q(H)$ as above.

If $\mathfrak{a} = \text{Lie}(A)$ with the exponential isomorphism

$$\exp : \mathfrak{a} \to A,$$

we define the **Gauss space** Gauss(A) to be the space corresponding to Gauss($\mathfrak{a}$) under the exponential map.

Let $\zeta \in \mathfrak{a}^\vee_{\mathbf{C}}$. For any function φ on A we put

$$\varphi_\zeta = \varphi\chi_\zeta \text{ that is } \varphi_\zeta(a) = \varphi(a)a^\zeta$$

or also additively

$$\varphi^{\mathfrak{a}}_\zeta(H) = \varphi^{\mathfrak{a}}(H)e^{\zeta(H)}.$$

Note that φ is in the Gauss space if and only if φ_ζ is in the Gauss space.

We recall that the additive version $\mathbf{M}_{\mathfrak{a}}$ of the Mellin transform is given by the integral

$$\mathbf{M}_{\mathfrak{a}}\varphi^{\mathfrak{a}}(\zeta) = \int_{\mathfrak{a}} \varphi^{\mathfrak{a}}(H)e^{\zeta(H)}dH \quad \text{for } \zeta \in \mathfrak{a}^\vee_{\mathbf{C}}.$$

For $\zeta = \xi + \mathbf{i}\lambda$ (with $\xi, \lambda \in \mathfrak{a}^\vee$), and $\eta \in \mathfrak{a}^\vee_{\mathbf{C}}$, we have

$$\mathbf{M}_{\mathfrak{a}}\varphi^{\mathfrak{a}}_\eta(\zeta) = (\mathbf{M}_{\mathfrak{a}}\varphi^{\mathfrak{a}})(\zeta + \eta) \quad \text{and especially} \quad \mathbf{M}_{\mathfrak{a}}\varphi^{\mathfrak{a}}(\zeta) = (\mathbf{M}_{\mathfrak{a}}\varphi^{\mathfrak{a}}_\xi)(\mathbf{i}\lambda).$$

Of course we have the corresponding functions defined on the multiplicative version A of $\mathfrak{a}$, for example $A = A_P$ as in §4. Then for a function φ on A, $\eta \in \mathfrak{a}^\vee_{\mathbf{C}}$,

$$(1) \qquad (\mathbf{M}\varphi_\eta)(\zeta) = \int_A \varphi(a)a^\eta a^\zeta da = (\mathbf{M}_{\mathfrak{a}}\varphi)(\eta + \zeta).$$

Mellin-Fourier inversion for φ in the Gauss space is

$$(2) \qquad \varphi(b) = \int_{\mathfrak{a}^\vee} (\mathbf{M}_{\mathfrak{a}}\varphi)(-\mathbf{i}\lambda)b^{\mathbf{i}\lambda}d\lambda,$$

or writing the translation,

$$(2\eta)\qquad \varphi(b) = \int_{\mathfrak{a}^\vee} (\mathbf{M}_{\mathfrak{a}}\varphi_{-\eta})(-\mathbf{i}\lambda)b^{\eta+\mathbf{i}\lambda}d\lambda \quad \text{for all } b \in A.$$

The Haar measures on $\mathfrak{a}$ and $\mathbf{i}\mathfrak{a}^\vee$ are assumed normalized so that no extra constant factor appears in the inversion formula. The second factor has been written so that instead of $\varphi_{-\eta}(b)$ on the left we have a factor b^η on the right inside the integral. For a further normalization, see below.

PROPOSITION 4.1. *Identifying $\mathfrak{a}$ with its dual under the positive definite scalar product, the Gauss space is self dual, i.e. the Mellin transform induces an isomorphism of* Gauss($\mathfrak{a}$) *with* Gauss($\mathfrak{a}$). *Or, without the identification, the Mellin transform gives an isomorphism*

$$\text{Gauss}(\mathfrak{a}) \to \text{Gauss}(\mathbf{i}\mathfrak{a}^\vee).$$

PROOF. This is just Fourier-Mellin inversion, combined with the self duality of the Gauss function $f(H) = e^{-H^2/2}$, and with the translation property (1). The scalar product $\langle H, H'\rangle$ on $\mathfrak{a}$ identifies $\mathfrak{a}$ with $\mathfrak{a}^\vee$, and the Fourier transform of this function f is $e^{-\lambda^2/2} = e^{(\mathbf{i}\lambda)^2/2}$. That $\mathbf{M}f_\eta$ is in the Gauss space then follows from (1). Note that it is convenient here (among other places) to work with the complexified space. For precise formulas, see **Mq 1** through **Mq 4** below.

PROPOSITION 4.2. *Let f be a bounded continuous function on $\mathfrak{a}$. Suppose that for every $H' \in \mathfrak{a}$ and $c > 0$, f is orthogonal to $e^{-c(H-H')^2}$. Then $f = 0$.*

PROOF. By assumption

$$0 = \int_{\mathfrak{a}} e^{-c(H-H')^2} f(H)dH.$$

Suppose we are dealing with the usual normalization of Haar measure on $\mathfrak{a}$. Let $c = 1/4t$, and multiply the above equation by $1/(4\pi t)^{r/2}$. The resulting expression is the convolution of the heat kernel with f. Letting $t \to 0$ and using the Dirac property of the heat kernel shows that $f(H') = 0$. This being true for every $H' \in \mathfrak{a}$ we conclude that $f = 0$, thus proving the proposition.

The above two general statements will suffice for most applications, but it will be useful at some point to have a more explicit version of Theorem 3.1. For convenience of reference, we tabulate the formulas in the present section. The given scalar product $\langle , \rangle_{\mathfrak{a}}$ on $\mathfrak{a}$ induces an isomorphism of $\mathfrak{a}$ with its dual

$$\mathfrak{a} \xrightarrow{\approx} \mathfrak{a}^\vee, \quad H \mapsto \lambda_H \quad \text{such that} \quad \lambda_H(H') = \langle H, H'\rangle_{\mathfrak{a}}.$$

This gives rise to the corresponding positive definite scalar product $\langle , \rangle_{\mathfrak{a}^\vee}$. For simplicity, we omit the indices. Note that this product on $\mathfrak{a}^\vee$ is positive definite. Having normalized the Haar measures on $\mathfrak{a}$ and $\mathfrak{a}^\vee$ to satisfy Fourier inversion, there is still a choice of scaling possible (multiply one of them by a positive constant, and the other by the inverse of the constant). We normalize still further. The euclidean volume form provides a natural Haar measure on each of $\mathfrak{a}$ and $\mathfrak{a}^\vee$. It is standard that multiplying these Haar measures by $(2\pi)^{-r/2}$ make them satisfy Fourier inversion. We can call this the **metric Fourier inversion normalization**, *which we now assume.*

Each one of the scalar products on $\mathfrak{a}$ and $\mathfrak{a}^\vee$ has a $\mathbf{C}$-bilinear extension to the complexifications $\mathfrak{a}_\mathbf{C}$ and $\mathfrak{a}_\mathbf{C}^\vee$ respectively. The extension to the imaginary axis $\mathbf{i}\mathfrak{a}^\vee$ is negative definite because for $\lambda, \mu \in \mathfrak{a}^\vee$,

$$\langle \mathbf{i}\lambda, \mathbf{i}\mu \rangle = -\langle \lambda, \mu \rangle.$$

We shall list some formulas which oil the mechanism of taking the Mellin transform of functions in the Gauss space. We let $q_0, q_0^\vee$ be the quadratic functions

$$q_0(H) = -\langle H, H' \rangle/2 \quad \text{and} \quad q_0^\vee(\zeta) = \langle \zeta, \zeta \rangle/2.$$

We have the standard fact of advanced calculus, expressing that e^{q_0} is self dual, under the metric Fourier inversion normalization. For simplicity of notation, we use $\mathbf{M}$ instead of $\mathbf{M}_\mathfrak{a}$ for the Mellin transform of function on $\mathfrak{a}$.

Mq 1. $\mathbf{M}(e^{q_0}) = e^{q_0^\vee}$.

Next we give formulas concerning scaling. Let

$$q_\mathfrak{a}(H) = -\langle H, H \rangle \quad \text{and} \quad q_{\mathbf{i}\mathfrak{a}^\vee}(\zeta) = \langle \zeta, \zeta \rangle.$$

Mq 2. Let $q = cq_0$ with $c > 0$. Then

$$\mathbf{M}(e^{cq_0}) = \frac{1}{c^{r/2}} e^{q_0^\vee / c}.$$

Mq 3. Let $q = cq_\mathfrak{a}$ with $c > 0$. Then

$$\mathbf{M}(e^{cq_\mathfrak{a}})(\zeta) = \frac{1}{(2c)^{r/2}} e^{\langle \zeta, \zeta \rangle / 4c}.$$

PROOF. Both formulas come from the definitions and a change of variables in the Mellin transform, letting $H \mapsto H/c^{\frac{1}{2}}$ and $H \mapsto H/(2c)^{\frac{1}{2}}$ respectively.

Mq 4. $\mathbf{M}(e^{cq_\mathfrak{a}+\eta})(\zeta) = (2c)^{-r/2} e^{\langle \zeta+\eta, \zeta+\eta \rangle / 4c} \quad$ for $\zeta, \eta \in \mathfrak{a}_\mathbf{C}^\vee$.

PROOF. Apply **Mq 3** and (1) (Mellin transform of $\varphi(H)e^{\eta(H)}$).

For some applications, we are given $\mu \in \mathfrak{a}^\vee$ and $t > 0$. We let

$$q^\vee(\zeta) = \langle \zeta, \zeta \rangle + \langle \zeta, \mu \rangle,$$

and we want to solve for the function φ such that

$$\mathbf{M}\varphi = e^{tq^\vee}.$$

The answer is given by:

PROPOSITION 4.3. *Let $q^\vee$ be as above. Let $t > 0$. Let*

$$q_t = \frac{1}{4t} q_{\mathfrak{a}} + \frac{\mu}{2} - \langle \mu, \mu \rangle \frac{t}{4}.$$

Then

$$\mathbf{M}(e^{q_t}) = (2t)^{r/2} e^{tq^\vee}.$$

PROOF. In **Mq 4**, let $c = 1/4t$ and $\eta = \mathbf{M}u/2$. Expanding out the square on the right side of **Mq 4** yields the formula of the proposition.

REMARK 1. *The above formulas hold under the metric Fourier inversion normalization.* Note that the factor $(2\pi)^{-r/2}$ on the metric (Riemannian) measures affects subsequent formulas. For example, with respect to the present normalization, the euclidean heat Gaussian on $\mathfrak{a}^\vee$ is given by

$$h_t(\lambda) = (2t)^{-r/2} e^{-\langle \lambda, \lambda \rangle / 4t}.$$

The usual additional factor $(2\pi)^{-r/2}$ is already taken care of by our normalization of Haar measures.

REMARK 2. The main point for us of the quadratic expression

$$q^\vee(-\zeta) = \langle \zeta, \zeta \rangle - \langle \zeta, \mu \rangle$$

is that with a suitably chosen μ, it becomes the eigenvalue of the Casimir operator on the character χ_ζ, Cf. Chapter 4, §3. Proposition 4.3 will be used in the considerations leading to Chapter 3 Theorem 4.1, which in turn will be relevant in the context of eigenvalues and solutions of the heat equation in Chapter 4.

We may now return to the Eisenstein series formed with other functions than characters, mentioned following Proposition 2.1.

PROPOSITION 4.4. *Let $\varphi \in \mathrm{Gauss}(A_P)$. Then the series*

$$E_P(\varphi) = \sum_{\Gamma_P \backslash \Gamma} \varphi((\gamma x)_{A_P})$$

converges absolutely. It is dominated by $E_P(\xi)$ for every $\xi > 2\rho_P$, meaning that for each x there is a finite subset $S(x) \subset \Gamma_P \backslash \Gamma$ such that for $\gamma \notin S(x)$ we have

$$|\varphi((\gamma x)_{A_P})| \leqq (\gamma x)_{A_P}^{\xi}.$$

PROOF. This is immediate from the quadratic exponential decay for φ as compared to linear exponential growth, since we know that the character Eisenstein series converges, by Theorem 1.1.

2.5. The parabolic Eisenstein integration formula

We note the inclusion of subgroups

$$\Gamma_P \subset \Gamma \subset G.$$

We shall deal with the integration of Eisenstein series over $\Gamma\backslash G$, so we meet naturally the integral

$$\int_{\Gamma_P\backslash G} f(x)dx = \int_{\Gamma\backslash G} \sum_{\gamma\in\Gamma_P\backslash\Gamma} f(\gamma x)dx$$

under conditions of absolute convergence. The next theorem expresses such an integral in terms of parabolic coordinates. Its proof will need a standard extension of the uniqueness of Haar measure to the case of a relatively invariant measure rather than an invariant measure. The statement is as follows.

LEMMA 5.1. *Let G be a locally compact group and H a closed subgroup. Let μ be a regular Borel measure on G/H, and let χ be a character on G such that $\mu(gS) = \chi(g)\mu(S)$ for all $g \in G$ and all measurable sets S. Then μ is uniquely determined up to a constant factor.*

PROOF. See Raghunathan [**Rag 72**], Chapter I, Lemma 1.4.

We may now formulate and prove the **parabolic Eisenstein integration formula**.

PROPOSITION 5.2. *Let $f \in C(\Gamma_P\backslash G)$ be right K-invariant and in $L^1(\Gamma_P\backslash G)$. Then up to a constant factor (normalizing Haar measures),*

$$\int_{\Gamma_P\backslash G} f(x)dx = \int_{\Gamma_{U_P}\backslash U_P} \int_{A_P} \int_{\Gamma_{G_P}\backslash G_P} f(uag)\delta_P^{-1}(a)dudgda. \tag{1}$$

PROOF. The coset space $P\backslash G$ is compact. Considering the inclusion

$$\Gamma_P \subset P \subset G,$$

we have, up to a normalizing constant factor,

$$\int_{\Gamma_P\backslash G} f(x)dx = \int_{P\backslash G} \int_{\Gamma_P\backslash P} f(px)dpdx_{P\backslash G} = \int_{\Gamma_P\backslash P} f(p)dp.$$

It will then suffice to prove that this functional on $C_c(\Gamma_P\backslash P)$ and the functional

$$f \mapsto \int_{\Gamma_{U_P}\backslash U_P} \int_{A_P} \int_{\Gamma_{G_P}\backslash G_P} f(uag)\delta_P^{-1}(a)dudgda$$

have the same character, namely δ_P when composed with right translations.

First note that the right side of the arrow is well-defined, i.e. the integrals are defined over the coset spaces as written. This is clear from the assumed Γ_P-invariance of f for the U_P-integral. For the G_P-integral, let $\gamma \in \Gamma_{G_P}$. Using that conjugation by γ leaves U_P stable, we obtain

$$f(ua\gamma g) = f(u\gamma ag) = f(\gamma u^{\gamma}.ag) = f(u^{\gamma}.ag).$$

The Haar measure on U_P is unchanged by conjugation with γ (because actually the homomorphism $g \mapsto |\det \mathbf{C}_\mathfrak{n}(g)|$ for $g \in G_P$ is trivial, since $\mathrm{Hom}(\mathrm{SL}_n(\mathbf{C}), \mathbf{R}_{>0})$ is trivial). Hence the du-integral is well defined over $\Gamma_{U_P}\backslash U_P$.

On P, the Haar measure transforms by δ_P under right translations. It is immediately verified that the triple integral transforms in the same way. Indeed, right translation by an element of G_P leaves the triple integral invariant since G_P is unimodular. Translating by an element b of A_P immediately shows that a factor $\delta_P(b)$ comes out. Since U_P is normal in P, a translation by an element u_1 on the right amounts to a translation by some conjugate of u_1 next to u, and then the fact that U_P is unimodular shows that this translation can be canceled, so we get the desired character under right translations. This concludes the proof.

CHAPTER 3

Adjointness and Inversion Relations

The preceding chapter dealt with the formalism of the trace, with respect to discrete subgroups. Next, we shall apply this formalism in connection with integration, especially the Fubini theorem relating integration on a group, subgroup and homogeneous space, and the parabolic coordinates integration formula. At the heart of the matter lies an adjointness relation between trace operators on the U_P-factor and the G_P-factor. Also involved is an inversion relation going back and forth between characters and functions in the Gauss space which can be used in Eisenstein series. We develop a basic formalism revolving around the characterization of cuspidal functions, i.e. functions ψ such that

$$\int_{\Gamma_{U_P}\backslash U_P} \psi(ux)du = 0 \quad \text{for all } x \in G.$$

We investigate variations of this property, involving convolutions with various functions, including Eisenstein series and the heat kernel.

In Chapter 4, we shall give applications of the formalism, showing for the first time the role of the analytic continuation of the heat Eisenstein series as functions of the variable $\zeta \in \mathfrak{a}^{\vee}_{P,\mathbf{C}}$ to the critical strip, and more specifically the center of the critical strip. With such analytic continuation, one gets insight into what amounts to one-parameter semigroups of anti-cuspidal projections.

3.1. Adjointness formulas and F-cuspidality

We start with some abstract nonsense which gives a general background for adjointness formulas involving integration.

Let G be a locally compact group and V, U two closed subgroups. We have inclusions

$$\begin{array}{ccccc} & & V & & \\ & \nearrow & & \searrow & \\ U_V = V_U = V \cap U & & & & G. \\ & \searrow & & \nearrow & \\ & & U & & \end{array}$$

We suppose all groups unimodular, so the coset spaces have invariant measures.

We shall define the two linear maps

$$T : \text{functions on } U\backslash G \longrightarrow \text{functions on } V\backslash G$$

$$T' : \text{functions on } V\backslash G \longrightarrow \text{functions on } U\backslash G$$

We leave the spaces of functions unspecified, what is needed is absolute convergence of the integrals we shall write down. The maps T, T' are defined by the formula

$$TF(x) = \int\limits_{V \cap U\backslash V} F(vx)dv \quad \text{and} \quad T'\psi(y) = \int\limits_{V \cap U\backslash U} \psi(uy)du.$$

In the application, V will be a discrete subgroup Γ and the integral involving v as variable will be a sum over $\gamma \in \Gamma$.

We have the integral scalar product on $V\backslash G$, defined by

$$[\psi_1, \psi_2]_{V\backslash G} = \int\limits_{V\backslash G} \psi_1(x)\psi_2(x)dx_{V\backslash G} \quad \text{for } \psi_1, \psi_2 \text{ on } V\backslash G;$$

and similarly on $U\backslash G$,

$$[F_1, F_2]_{U\backslash G} = \int\limits_{U\backslash G} F_1(y)F_2(y)dy_{U\backslash G} \quad \text{for } F_1, F_2 \text{ on } U\backslash G.$$

Under conditions guaranteeing absolute convergence, we have the **adjointness relation**

$$[TF, \psi]_{V\backslash G} = [F, T'\psi]_{U\backslash G}.$$

PROOF. The formal computation is

$$\begin{aligned}
[TF, \psi]_{V\backslash G} &= \int\limits_{V\backslash G} TF(x)\psi(x)dx_{V\backslash G} \\
&= \int\limits_{V\backslash G}\int\limits_{V\cap U\backslash V} F(vx)\psi(x)dvdx_{V\backslash G} \\
&= \int\limits_{V\cap U\backslash G} F(x)\psi(x)dx_{V\cap U\backslash G}
\end{aligned}$$

which is now in a form symmetric between V and U. Thus we can rewind this integral just as above, to see that it is

$$= [F, T'\psi]_{U\backslash G},$$

thus finishing the formal proof.

For example, if $F \in C_c(U\backslash G)$ and $\psi \in L^2(V\backslash G)$, then the formal argument is valid. We shall give other conditions later designed for the applications.

In practice, the situation will be slightly more complicated because there is still a third variable, so we shall give the proof in a self contained way to take this additional structure into account. It comes from the parabolic coordinates, of which there are three, respectively in U_P, A_P, and G_P/K_{G_P}.

Application to Eisenstein series.

We shall use the adjointness formula to get information about cuspidality. So we deal with our usual $G = \mathrm{SL}_n(\mathbf{C})$ and a reduced parabolic $P = U_P A_P G_P$. The adjointness formulas are concerned with the discrete group on the left. We have the inclusions

$$\Gamma_{U_P} \subset \Gamma_P \subset \Gamma \subset G.$$

Note that the elements of Γ_{G_P} form a natural set of representatives for the cosets of $\Gamma_{U_P}\backslash\Gamma_P$. We denote such cosets or their representatives by

$$\{\eta\} = \Gamma_{U_P}\backslash\Gamma_P = \Gamma_{G_P}.$$

Suppose that $f \in L^1(\Gamma_{U_P}\backslash G)$ and f is continuous right K-invariant. Then

$$(1) \qquad \int_{\Gamma_{U_P}\backslash G} f(x)dx = \int_{\Gamma\backslash G} \int_{\Gamma_P\backslash\Gamma} \int_{\Gamma_{U_P}\backslash\Gamma_P} f(\eta\gamma x)d\eta d\gamma dx_{\Gamma\backslash G}.$$

The inner integrals are of course sums since they are taken over discrete coset spaces of discrete groups, and so (1) may be rewritten

$$(2a) \qquad \int_{\Gamma_{U_P}\backslash G} f(x)dx = \int_{\Gamma\backslash G} \sum_{\gamma\in\Gamma_P\backslash\Gamma} \sum_{\eta\in\Gamma_{U_P}\backslash\Gamma_P} f(\eta\gamma x)dx_{\Gamma\backslash G}$$

$$= \int_{\Gamma\backslash G} \sum_{\gamma\in\Gamma_P\backslash\Gamma} \mathrm{Tr}_{\Gamma_{G_P}}(f)(\gamma x)dx_{\Gamma\backslash G}.$$

We shall deal with different types of functions f decomposed into products. Some factors will be characters, and other factors are designed to ensure convergence of the integrals involved. We now give another expression for the above integral. We define the P-**cuspidal integral** or P-**cuspidal trace** to be

$$\mathrm{Tr}_{\Gamma_{U_P}\backslash U_P}(f)(x) = \int_{\Gamma_{U_P}\backslash U_P} f(ux)du.$$

This integral will play a central, independent role, starting with §2 below.

From the measure isomorphism $P/K_{G_P} \to G/K$ and the parabolic coordinates integration formula, Chapter 1, §7 (5), we get the formula

(2b)
$$\int_{\Gamma_{U_P}\backslash G} f(x)dx = \int_{\Gamma_{U_P}\backslash P} f(p)dp$$
$$= \int_{\Gamma_{U_P}\backslash U_P}\int_{A_P}\int_{G_P} f(uag)\delta_P^{-1}(a)dgdadu$$
$$= \int_{A_P}\int_{G_P} ((\mathrm{Tr}_{\Gamma_{U_P}\backslash U_P})(f))(ag)\delta_P^{-1}(a)dgda.$$

Putting (2a) and (2b) together yields the **Eisenstein parabolic integration formula**

(3)
$$\int_{\Gamma\backslash G} \sum_{\gamma\in\Gamma_P\backslash\Gamma} \mathrm{Tr}_{\Gamma_{G_P}}(f)(\gamma x)dx_{\Gamma\backslash G}$$
$$= \int_{A_P}\int_{G_P} ((\mathrm{Tr}_{\Gamma_{U_P}\backslash U_P})(f))(ag)\delta_P^{-1}(a)dgda.$$

This identity will be used over and over again in this chapter. Note that

$$U_P\backslash P = U_P\backslash U_PA_PG_P = A_PG_P.$$

So the parabolic coordinates have the effect of shifting the $\Gamma\backslash G$ integral to $U_P\backslash P = A_PG_P$.

We shall now deal with adjointness formulas.

Let $F_0 \in \mathrm{BC}(\Gamma_{G_P}\backslash G_P/K_{G_P})$, $\varphi \in C(A_P)$. We recall the definition of the **Eisenstein series** $E_P(F_0,\varphi)$, which at $x \in G$ has the value

$$E_P(F_0,\varphi)(x) = \sum_{\gamma\in\Gamma_P\backslash\Gamma} F_0((\gamma x)_{\mathbf{x}_{G_P}})\varphi((\gamma x)_{A_P}),$$

under conditions on F_0, φ when the series is absolutely convergent. Note that

$$(F_0,\varphi) \mapsto E_P(F_0,\varphi)$$

is bilinear, and it has the formal properties of a non-singular bilinear form. We shall mostly deal with the case when $F_0 = \mathrm{Tr}_{\Gamma_{G_P}}(F)$ for some function F on G_P/K_{G_P}.

We define the P-**cuspidal operator**

$$\mathrm{Tr}_{\Gamma_{U_P}\backslash U_P} : C(\Gamma\backslash G/K) \to C(\Gamma_{G_P}\backslash A_PG_P/K_{G_P}),$$

acting on functions $\psi \in C(\Gamma\backslash G/K)$, namely we define the **cuspidal trace**

$$\mathrm{Tr}_{\Gamma_{U_P}\backslash U_P}(\psi)(ag) = \int_{\Gamma_{U_P}\backslash U_P} \psi(uag)du.$$

The rest of this chapter will be concerned with this operator and its kernel (in the sense of linear algebra).

The symmetric scalar product of a function E with a function $\psi \in C(\Gamma\backslash G/K)$ (so ψ is left Γ-invariant and right K-invariant), is defined by

$$[E, \psi]_{\Gamma\backslash G} = \int\limits_{\Gamma\backslash G} E(x)\psi(x) dx_{\Gamma\backslash G}$$

whenever absolutely convergent.

The next theorem will be called the **Eisenstein adjointness theorem**. Note that on the upper half plane, when $U = U_P$, $A = A_P$, and $G_P = \{1\}$, the first part amounts to what is sometimes called the **Rankin-Selberg** method when φ is a character χ_ζ. Cf. for instance the exposition in Zagier [**Zag 79**], Proposition 2, p. 314. In words, the scalar product of a function on $\Gamma\backslash G/K$ with a character Eisenstein series equals the Mellin transform of the constant term in the Fourier expansion.

THEOREM 1.1. **First Part.** *Let $F_0 \in \mathrm{BC}(\Gamma_{G_P}\backslash G_P/K_{G_P})$, $\varphi \in C(A_P)$, $\psi \in BC(\Gamma\backslash G/K)$. Assume that*

$$E_P(|F_0|, |\varphi|) \in L^1(\Gamma\backslash G).$$

Then with the measure $\delta_P^{-1}(a)dadg$ on $A_P G_P$,

$$[E_P(F_0, \varphi), \psi]_{\Gamma\backslash G} = [F_0\varphi, \mathrm{Tr}_{\Gamma_{U_P}\backslash U_P}(\psi)]_{\Gamma_{G_P}\backslash A_P G_P}.$$

Second Part. *Let $F \in C(G_P/K_{G_P})$, $\varphi \in C(A_P)$, $\mathrm{Tr}_{\Gamma_{G_P}}(|F|)$ bounded, and*

$$E_P(\mathrm{Tr}_{\Gamma_{G_P}}(|F|, |\varphi|) \in L^1(\Gamma\backslash G).$$

Let $(F\varphi)(ag) = F(g)\varphi(a)$. Let $\psi \in BC(\Gamma\backslash G/K)$. Then

$$[E_P(\mathrm{Tr}_{\Gamma_{G_P}}(F), \varphi)), \psi]_{\Gamma\backslash G} = [F\varphi, \mathrm{Tr}_{\Gamma_{U_P}\backslash U_P}(\psi))]_{A_P G_P}$$

$$\text{(i)} \qquad = \int\limits_{A_P}\int\limits_{G_P} F(g)\mathrm{Tr}_{\Gamma_{U_P}\backslash U_P}(\psi)(ag)\varphi(a)\delta_P^{-1}(a)dgda$$

$$\text{(ii)} \qquad = \int\limits_{A_P}\int\limits_{\Gamma_{G_P}\backslash G_P} \mathrm{Tr}_{\Gamma_{G_P}}(F)(g)\mathrm{Tr}_{\Gamma_{U_P}\backslash U_P}(\psi)(ag)\varphi(a)\delta_P^{-1}(a)dgda$$

PROOF. We leave the first part to the reader (see Chapter 2, Proposition 5.2), and prove the second part. From the definitions,

$$\int\limits_{\Gamma\backslash G}\left[\sum_{\gamma\in\Gamma_P\backslash\Gamma} \mathrm{Tr}_{\Gamma_{G_P}}(F)((\gamma x)_{\mathbf{x}_{G_P}}\varphi((\gamma x)_{A_P}))\right]\psi(x)dx_{\Gamma\backslash G}$$

$$= \int\limits_{\Gamma\backslash G}\left[\sum_{\gamma\in\Gamma_P\backslash\Gamma} \mathrm{Tr}_{\Gamma_{G_P}}(F)((\gamma x)_{\mathbf{x}_{G_P}})\varphi((\gamma x)_{A_P}))\psi(\gamma x)\right]dx_{\Gamma\backslash G}$$

to which we can apply (3) to conclude the proof of (i). The formulation (ii) is trivially equivalent to (i), so the proof is done.

REMARK. The integral of Theorem 1.1 (i) will occur again, in an expanded form. In light of the expanded definition of $C_{P,\eta}$ in §3(3) below, we denote the present integral by

$$C_0(F,\psi,\varphi) = \int_{A_P}\int_{G_P} F(g)\mathrm{Tr}_{\Gamma_{U_P}\backslash U_P}(\psi)(ag)\varphi(a)\delta_P^{-1}(a)dgda.$$

It is trilinear in F, ψ, φ.

We consider quadruples

$$(\xi, \varphi, F, \psi)$$

with $\xi \in \mathfrak{a}_P^\vee$, $\varphi \in C(A_P)$, $F \in C(G_P/K_{G_P})$ and $\psi \in C(\Gamma\backslash G/K)$. We say such a quadruple is P-**admissible** under the following conditions:

P-**ADM 1.** $\xi > 2\rho_P$ so Theorems 1.1 and 1.4 of Chapter 2 are valid;

P-**ADM 2.** $\varphi \in \mathrm{Gauss}(A_P)$;

P-**ADM 3.** $\mathrm{Tr}_{\Gamma_{G_P}}(|F|)$ is bounded; we let

$$F_0 = \mathrm{Tr}_{\Gamma_{G_P}}(F);$$

P-**ADM 4.** $\psi E_P(|F|, \chi_\xi) \in L^1(\Gamma\backslash G)$, and ψ is bounded.

By P-**ADM 4**, we may take the scalar product. As in Chapter 2, §3, we let $\zeta = \xi + \mathbf{i}\lambda \in \mathfrak{a}_{P,\mathbf{C}}^\vee$ and

$$[E_P(F_0,\chi_\zeta),\psi]_{\Gamma\backslash G} = \int_{\Gamma\backslash G} E_P(F_0,\chi_\zeta)(x)\psi(x)dx_{\Gamma\backslash G}.$$

Next comes what we call the **Eisenstein-Mellin inversion theorem.**

THEOREM 1.2. *Let the quadruple above be P-admissible. Let ζ be the variable in $\mathfrak{a}_{P,\mathbf{C}}^\vee$. Then for $x \in G$ we have*

(i)
$$E_P(F_0,\varphi)(x) = \int_{\mathrm{Re}(\zeta)=\xi} E_P(F_0,\chi_\zeta)(x)(\mathbf{M}_{\mathfrak{a}}\varphi)(-\zeta)d\mathrm{Im}(\zeta)$$

(ii)
$$[E_P(F_0,\varphi),\psi]_{\Gamma\backslash G} = \int_{\mathrm{Re}(\zeta)=\xi} [E_P(F_0,\chi_\zeta),\psi]_{\Gamma\backslash G}(\mathbf{M}_{\mathfrak{a}}\varphi)(-\zeta)d\mathrm{Im}(\zeta)$$

Here $\mathbf{M}_{\mathfrak{a}}$ denote the Mellin transform from $\mathrm{Gauss}(A_P)$ to functions on $\mathfrak{a}_{P,\mathbf{C}}^\vee$. For the most part, we omit the subscript $\mathfrak{a}$, and let the context determine the domain and range of the various functions (multiplicative or additive).

PROOF. We use the Mellin-Fourier inversion of Chapter 2, §4,(2). For each γ we put $b = (\gamma x)_{A_P}$. We multiply the inversion formula by

$$F_0((\gamma x)\mathbf{x}_{G_P}).$$

The factor $F_0((\gamma x)\mathbf{x}_{G_P})$ of course acts as a constant with respect to the integral. By the admissibility assumptions, the sum over $\gamma \in \Gamma_P\backslash\Gamma$ can be taken under the integral sign, and yields precisely the expression on the left in the theorem. This concludes the proof.

At some point we shall need the following formulation using Chapter 2, §4, (2η).

THEOREM 1.2η. *Let the quadruple above be P-admissible. Let $\eta \in \mathfrak{a}^{\vee}_{P,\mathbf{C}}$. Abbreviate $F_0 = \mathrm{Tr}_{\Gamma_{G_P}}(F)$. Then for $\mathrm{Re}(\eta) = \xi$ and $x \in G$, we have*

$$\text{(i)} \qquad E_P(F_0, \varphi)(x) = \int_{\mathfrak{a}^{\vee}_P} E_P(F_0, \chi_{\eta+\mathbf{i}\lambda})(x)(\mathbf{M}_{\mathfrak{a}}\varphi)(-\eta - \mathbf{i}\lambda)d\lambda$$

and

$$\text{(ii)} \qquad [E_P(F_0, \varphi), \psi]_{\Gamma\backslash G} = \int_{\mathfrak{a}^{\vee}_P} [E_P(F_0, \chi_{\eta+\mathbf{i}\lambda}), \psi]_{\Gamma\backslash G}(\mathbf{M}_{\mathfrak{a}}\varphi)(-\eta - \mathbf{i}\lambda)d\lambda.$$

We remind the reader that

$$\varphi_{-\eta}(a) = \varphi(a)a^{-\eta} \quad \text{and} \quad (\mathbf{M}_{\mathfrak{a}}\varphi_{-\eta})(\zeta) = (\mathbf{M}_{\mathfrak{a}}\varphi)(-\eta + \zeta).$$

We shall immediately give two applications of the adjointness and inversion theorems. The first is to F-cuspidality and the second is to showing that a "constant term" is equal to 0.

F-cuspidality

The next theorem will allow us to define the notion of (P, F)-cuspidality by any one of three equivalent conditions.

THEOREM 1.3. *Assume that ξ, F, ψ satisfy P-**ADM 1**, P-**ADM 3**, P-**ADM 4** respectively. Let $F_0 = \mathrm{Tr}_{\Gamma_{G_P}}(F)$. The following conditions are equivalent:*

(i) $E_P(F_0, \chi_\zeta) \perp \psi$ *on* $\Gamma\backslash G$ *for all* ζ *with* $\mathrm{Re}(\zeta) = \xi$, *that is*

$$[E_P(F_0, \chi_\zeta), \psi]_{\Gamma\backslash G} = 0.$$

(ii) $E_P(F_0, \varphi) \perp \psi$ *on* $\Gamma\backslash G$ *for all* $\varphi \in \mathrm{Gauss}(A_P)$, *that is,*

$$[E_P(F_0, \varphi), \psi]_{\Gamma\backslash G} = 0.$$

(iii) $F \perp \mathrm{Tr}_{\Gamma_{U_P}\backslash U_P}(\psi) \circ a$ *on* G_P *for all* $a \in A_P$, *that is*

$$[F, \mathrm{Tr}_{\Gamma_{U_P}\backslash U_P}(\psi) \circ a]_{G_P} = \int_{G_P} F(g)\mathrm{Tr}_{\Gamma_{U_P}\backslash U_P}(\psi)(ag)dg = 0.$$

PROOF. Assume (i). By Theorem 1.2(ii), we get

$$[E_P(F_0, \varphi), \psi]_{\Gamma \backslash G} = 0 \quad \text{for all } \varphi \in \text{Gauss}(A_P),$$

which is (ii). Assume (ii). Let

$$h_1(a) = \int_{G_P} F(g) \text{Tr}_{\Gamma_{U_P} \backslash U_P}(\psi)(ag) dg.$$

By assumptions P-**ADM 3** and P-**ADM 4**, h_1 is bounded continuous. Note that the map $\varphi \mapsto \varphi\delta_P^{-1}$ is a bijection of the Gauss space with itself. Hence by Theorem 1.1, h_1 is orthogonal to the Gauss functions, whence $h_1 = 0$ by Proposition 4.2 of Chapter 2. This concludes the proof that (i) $\Longrightarrow$ (ii) $\Longrightarrow$ (iii).

Conversely, the steps are reversible. Assume (iii). By Theorem 1.1 we get

$$[E_P(F_0, \varphi), \psi]_{\Gamma \backslash G} = 0 \quad \text{for all } \varphi \in \text{Gauss}(A_P).$$

In Theorem 1.2(ii), the right side is 0. But the Mellin transform gives an isomorphism $\text{Gauss}(A_P) \to \text{Gauss}(\mathbf{i}\mathfrak{a}_P^\vee)$, so by Proposition 4.2 of Chapter 2 again,

$$[E_P(F_0, \chi_\zeta), \psi]_{\Gamma \backslash G} = 0,$$

thus proving (i), and concluding the proof.

Remark. Note that in Theorem 1.3, we fixed $\xi > 2\rho_P$. From the conditions where ξ does not appear, it follows that the conditions are independent of ξ satisfying $\xi > 2\rho_P$.

Fix the function F in Theorem 1.3. We define a function ψ to be (P, F)-**cuspidal** if the three orthogonality conditions of Theorem 1.3 are satisfied. However, this notion of (P, F)-cuspidality, even applied to various families of functions F, is not the one which leads to the simplest spectral decomposition. We gave Theorems 1.1 and 1.3 partly to illustrate in a simpler case arguments which will be given in the next two sections, involving the heat kernel which requires two variables instead of one. Thus we shall have to repeat the arguments from the proofs of Theorems 1.1 and 1.3 in a slightly more complicated context. The pay off will come from having subsequently a much simpler formal framework for the spectral decomposition. In §2, we deal with the notion of P-cuspidality rather than (P, F)-cuspidality. In §3, we are concerned with initial conditions, which will be used in connection with the heat equation in the next chapter.

Vanishing of a "constant term"

We return to the source, namely Theorem 1.2(i), which we shall apply to get directly a result involving the integral over $\mathbf{i}\mathfrak{a}_P^\vee$. Again let $F_0 = \text{Tr}_{\Gamma_{G_P}}(F)$.

THEOREM 1.4. *Let $\xi = \text{Re}(\zeta)$ and $\xi > 2\rho_P$. Suppose $|F|_0 = \text{Tr}_{\Gamma_{G_P}}(|F|)$ bounded and $\zeta \mapsto E_P(F_0, \chi_\zeta)(x)$ in $L^1(\xi + \mathbf{i}\mathfrak{a}_P^\vee)$ for each $x \in G$. Then*

$$\int_{\text{Re}(\zeta)=\xi} E_P(F_0, \chi_\zeta)(x) d\text{Im}(\zeta) = 0 \quad \textit{for all } x \in G.$$

In other words, $E_P(F_0, \chi_\zeta)$ is orthogonal to the constants on the vertical axis $\xi + \mathbf{i}\mathfrak{a}_P^\vee$.

PROOF. By Theorem 1.2(i), for φ in the Gauss space,

$$(*) \qquad \int\limits_{\mathrm{Re}(\zeta)=\xi} E_P(F_0, \chi_\zeta)\mathbf{M}_\mathfrak{a}\varphi(-\zeta)d\mathrm{Im}(\zeta) = E_P(F_0, \varphi)(x).$$

We now need a family $\{\varphi_c\}$ of functions in Gauss(A_P) indexed by the positive reals, and satisfying the limits:

$$\lim_{c\to\infty} \mathbf{M}_\mathfrak{a}\varphi_c = 1 \quad \text{but} \quad \lim_{c\to\infty} \varphi_c = 0,$$

in a way that we can apply the dominated convergence theorem. We let

$$\varphi_c^\mathfrak{a}(H) = (2c)^{r/2}e^{-cH^2}, \quad \text{where} \quad H^2 = \langle H, H\rangle.$$

By Chapter 2, §4, formula **Mq 3**, we know that

$$(\mathbf{M}_\mathfrak{a}\varphi_c)(-\zeta) = e^{\langle\zeta,\zeta\rangle/2c}.$$

Then we have at least the first limit, that the exponential term inside the integral (*) approaches 1, and is bounded. We can apply the dominated convergence theorem to get the desired integral on the left side.

On the other hand, the right side of (*) can be written in full as

$$E_P(F_0, \varphi_c)(x) = \sum_{\gamma\in\Gamma_P\backslash\Gamma} \mathrm{Tr}_{\Gamma_{G_P}}(F)((\gamma x)\mathbf{x}_{G_P}(2c)^{r/2}e^{-c(H(\gamma x))^2},$$

where $H(\gamma x) = \log(\gamma x)_{A_P}$. For each $\gamma \in \Gamma_P\backslash\Gamma$, the set of $x \in G/K$ such that

$$(H(\gamma x))^2 = 0$$

is a proper real analytic subset, and so is the complement of an open dense set. Taking the union over all γ, the set of $x \in G/K$ such that there exists $\gamma \in \Gamma_P\backslash\Gamma$ for which $(H(\gamma x))^2 = 0$ is the complement of a dense set S. For $x \in S$, the series converges absolutely by Chapter 2, Proposition 4.4, and commutes with the limit as $c \to \infty$ by dominated convergence. This limit is 0 for each term, whence also for the series. This concludes the proof.

REMARK. The integral of Theorem 1.4 will recur frequently. We call it the **constant term integral**, and denote it by I_ξ. Thus for a function $f(\xi)$, we **define**

$$I_\xi f = \int\limits_{\mathrm{Re}(\zeta)=\xi} f(\zeta)d\mathrm{Im}(\zeta).$$

In other notation and normalization, it is the Fourier transform at 0.

Introduction to the next sections

The next two sections run a parallel course. As we saw in Chapter 2, we can deal with an Eisenstein series in two variables ζ_1, ζ_2 or with the case of $\zeta, \bar{\zeta}$. The two sections consider the adjointness and inversion relations in each one of these two cases. The main result for the $\zeta, \bar{\zeta}$ case concerns a heat Eisenstein characterization of P-cuspidality. The parallel result using the two separate variables ζ_1, ζ_2 will be combined with the study of the heat equation in Chapter 4, §3 and §4, getting to the heart of the heat equation, i.e. the uniqueness of solutions with given initial condition. These give rise to a cuspidal operator which we denote by $J_{P,\xi,t}$. The first formulas concerning this operator are given in the present chapter because they depend only on the basic formalism of this chapter, and the Dirac property of the heat kernel.

3.2. Adjointness and initial condition formulas

In this section, we record formulas in the context of the two-variable heat Eisenstein series. We thus go back directly to §1, (3), namely the Eisenstein parabolic integration formula. In particular, we get the initial condition formula for the two variable heat Eisenstein series.

Let $\mathbf{K}_{\mathbf{X}_{G_P}}$ be the heat kernel on $\mathbf{X}_{G_P} = G_P/K_P$. Let $\psi \in C(\Gamma \backslash G/K)$ and $\zeta_2 \in \mathfrak{a}^{\vee}_{P,\mathbf{C}}$. In §1, (3) let

$$(1) \qquad f(y) = f_{t,x,\zeta_2}(y) = \mathrm{Tr}_{\Gamma_{G_P}}(\mathbf{K}_{\mathbf{X}_{G_P}})(t, x_{\mathbf{X}_{G_P}}, y_{\mathbf{X}_{G_P}}) y_{A_P}^{\zeta_2} \psi(y).$$

We view ζ_2 now as fixed, $\mathrm{Re}(\zeta_2) = \xi > 2\rho_P$ so that the Eisenstein series $E_P(\chi_{\zeta_2})$ is absolutely convergent (cf. Chapter 2, Theorems 1.1 and 1.3). The above function f depends on the three parameters t, x, ζ_2. As in Chapter 2, §3 we define

$$\begin{aligned} &E_P^{(2)}(\mathrm{Tr}_{\Gamma_{G_P}}(\mathbf{K}_{\mathbf{X}_{G_P}}), \chi_{\zeta_2})(t, x, y) \\ &\qquad = \sum_{\gamma \in \Gamma_P \backslash \Gamma} \mathrm{Tr}_{\Gamma_{G_P}}(\mathbf{K}_{\mathbf{X}_{G_P}})(t, x_{\mathbf{X}_{G_P}}, (\gamma y)_{\mathbf{X}_{G_P}})(\gamma y)_{A_P}^{\zeta_2}. \end{aligned}$$

Thus $E_P^{(2)}(\mathrm{Tr}_{\Gamma_{G_P}}(\mathbf{K}_{\mathbf{X}_{G_P}}), \chi_{\zeta_2})$ is just an F-Eisenstein series, with

$$F(y) = F_{t,x}(y) = \mathbf{K}_{\mathbf{X}_{G_P}}(t, x_{\mathbf{X}_{G_P}}, y_{\mathbf{X}_{G_P}}).$$

Leaving out the variables, we have

$$E_P^{(2)}(\mathrm{Tr}_{\Gamma_{G_P}}(\mathbf{K}_{\mathbf{X}_{G_P}}), \chi_{\zeta_2})\psi = \mathrm{Tr}_{\Gamma_P \backslash \Gamma}(f).$$

PROPOSITION 2.1. *Assume that $E_P^{(2)}(\mathbf{K}_{\mathbf{X}_{G_P}}, \chi_{\zeta_2})(t, x, y)\psi(y)$ as a function of y with (t, x) fixed is in $L^1(\Gamma \backslash G/K)$. Then we have the identity*

$$\begin{aligned} &(E_P^{(2)}(\mathrm{Tr}_{\Gamma_{G_P}}(\mathbf{K}_{\mathbf{X}_{G_P}}), \chi_{\zeta_2}) * \psi)(t, x, y) \\ &\qquad = \int_{\Gamma \backslash G} E_P^{(2)}(\mathrm{Tr}_{\Gamma_{G_P}}(\mathbf{K}_{\mathbf{X}_{G_P}}), \chi_{\zeta_2})(t, x, y)\psi(y)dy \end{aligned}$$

$$(2) \qquad = \int_{\Gamma_{U_P} \backslash G} \mathbf{K}_{\mathbf{X}_{G_P}}(t, x_{\mathbf{X}_{G_P}}, y_{\mathbf{X}_{G_P}})\psi(y) y_{A_P}^{\zeta_2} dy$$

$$= \int_{A_P} \int_{G_P} \mathbf{K}_{\mathbf{X}_{G_P}}(t, x_{\mathbf{X}_{G_P}}, g) \mathrm{Tr}_{\Gamma_{U_P} \backslash U_P}(\psi)(ag) a^{\zeta_2 - 2\rho_P} dg da. \tag{3}$$

Note that $\psi(\gamma y) = \psi(y)$ *by assumption.*

PROOF. Special case of §1, (3), the fundamental Eisenstein integration identity.

Let $\psi \in C(\Gamma\backslash G/K)$. We say that ψ has P-**exponential quadratic decay** if there exists $\varphi \in \mathrm{Gauss}(A_P)$ such that for all $y = uag$ in P-parabolic coordinates, we have

$$|\psi(uag)| \leqq \varphi(a).$$

PROPOSITION 2.2. *Let* $\psi \in C(\Gamma\backslash G/K)$ *have* P*-exponential quadratic decay. Suppose that* $\mathrm{Re}(\zeta_2) > 2\rho_P$*. Then*

$$\lim_{t\to 0} (E_P^{(2)}(\mathrm{Tr}_{\Gamma_{G_P}}(\mathbf{K}_{\mathbf{X}_{G_P}}), \chi_{\zeta_2}) * \psi)(t, x) = \int_{A_P} \mathrm{Tr}_{\Gamma_{U_P} \backslash U_P}(\psi)(a x_{\mathbf{X}_{G_P}}) a^{\zeta_2 - 2\rho_P} da.$$

PROOF. This is immediate from Proposition 2.1, interchanging the integrals over A_P and G_P, and using the Dirac property of the heat kernel.

As in Chapter 2, §2 (5), the Eisenstein series $E_P(\mathrm{Tr}_{\Gamma_{G_P}}(\mathbf{K}_{\mathbf{X}_{G_P}}), \chi_{\zeta_1}, \chi_{\zeta_2})$ can be written as an iterated Eisenstein series, namely

$$E_P(\mathrm{Tr}_{\Gamma_{G_P}}(\mathbf{K}_{\mathbf{X}_{G_P}}), \chi_{\zeta_1}, \chi_{\zeta_2}) = E_P^{(1)}(E_P^{(2)}(\mathrm{Tr}_{\Gamma_{G_P}}(\mathbf{K}_{\mathbf{X}_{G_P}}), \chi_{\zeta_2}), \chi_{\zeta_1}). \tag{4}$$

Therefore

$$\begin{aligned} &E_P(\mathrm{Tr}_{\Gamma_{G_P}}(\mathbf{K}_{\mathbf{X}_{G_P}}), \chi_{\zeta_1}, \chi_{\zeta_2}) * \psi)(t, x) \\ &\quad = E_P^{(1)}(E_P^{(2)}(\mathrm{Tr}_{\Gamma_{G_P}}(\mathbf{K}_{\mathbf{X}_{G_P}}), \chi_{\zeta_2}) * \psi, \chi_{\zeta_1})(t, x). \end{aligned} \tag{5}$$

The convolution is of course the scalar product on $\Gamma\backslash G$ with respect to the y-variable. By Proposition 2.1 the expression (5) can be written in full in the form

$$\begin{aligned} &(E_{P,\Gamma,\mathbf{K}}(\chi_{\zeta_1}, \chi_{\zeta_2}) * \psi)(t, x) = \\ &\sum_{\gamma_1 \in \Gamma_P \backslash \Gamma} (\gamma_1 x)_{A_P}^{\zeta_1} \int_{A_P} \int_{G_P} \mathbf{K}_{\mathbf{X}_{G_P}}(t, (\gamma_1 x)_{\mathbf{X}_{G_P}}, g) \mathrm{Tr}_{\Gamma_{U_P} \backslash U_P}(\psi)(ag) a^{\zeta_2 - 2\rho_P} dg da. \end{aligned} \tag{6}$$

Taking the limit under the integral sign, and using Proposition 2.2 yields:

PROPOSITION 2.3. *Under the assumptions of Proposition 2.2,*

$$\begin{aligned} &\lim_{t\to 0} E_P(\mathrm{Tr}_{\Gamma_{G_P}}(\mathbf{K}_{\mathbf{X}_{G_P}}), \chi_{\zeta_1}, \chi_{\zeta_2}) * \psi)(t, x) \\ &\quad = \sum_{\gamma \in \Gamma_P \backslash \Gamma} (\gamma x)_{A_P}^{\zeta_1} \int_{A_P} \mathrm{Tr}_{\Gamma_{U_P} \backslash U_P}(\psi)(a(\gamma x)_{\mathbf{X}_{G_P}}) a^{\zeta_2 - 2\rho_P} da. \end{aligned}$$

REMARK. In some applications, we shall take $\psi = E_P(F, \varphi)$ with appropriate functions F, φ.

3.3. P-cuspidality and heat Eisenstein series

As before, we let $\mathbf{X} = G/K$ with $G = \mathrm{SL}_n(\mathbf{C})$, $\Gamma = \mathrm{SL}_n(\mathbf{Z}[\mathbf{i}])$. Let $\mathbf{K} = \mathbf{K}_{\mathbf{X}}$ be the heat kernel on $\mathbf{X}$. The Gangolli formula gives us all the estimates we need to make the following arguments valid. By Chapter 2, Proposition 2.4 and the twisted Fubini theorem for the subgroup $\Gamma \subset G$, we conclude that $\mathrm{Tr}_\Gamma(\mathbf{K})$ is in $L^1(\Gamma\backslash G)$ with respect to each variable x, y, and its total integral is 1.

Let $\psi \in \mathrm{BC}(\Gamma\backslash G/K)$, meaning ψ is bounded continuous on G, left Γ-invariant and right K-invariant. Let P be a standard reduced parabolic subgroup. We defined the **(cuspidal) trace** $\mathrm{Tr}_{\Gamma_{U_P}\backslash U_P}(\psi)$ as a function on G or G/K by

$$\mathrm{Tr}_{\Gamma_{U_P}\backslash U_P}(\psi)(x) = \int\limits_{\Gamma_{U_P}\backslash U_P} \psi(ux)du.$$

Note that $\mathrm{Tr}_{\Gamma_{U_P}\backslash U_P}(\psi)$ was then viewed as a function on $A_P G_P/K_P$ by Chapter 1, Theorem 4.1. We define ψ to be P-**cuspidal** if the above trace is 0, that is, for all $x \in G$ (or $x \in G/K$ because of the right K-invariance) we have

$$\int\limits_{\Gamma_{U_P}\backslash U_P} \psi(ux)du = 0.$$

We shall establish conditions of cuspidality in terms of the heat kernel, and the heat Eisenstein series. These are contained in Theorem 3.1 and Theorem 3.5. The next theorem is the simplest manifestation of these conditions, and already gives the flavor of the more substantial results to come. It is valuable because it admits a very short proof, devoid of further technicalities.

We shall deal with convolutions. First with the trace of the heat kernel itself, $\mathrm{Tr}_\Gamma(\mathbf{K})(t, x, y)$, which is a function of two variables on $\mathbf{X} = G/K$. For a function $\psi \in \mathrm{BC}(\Gamma\backslash G/K)$, and a function $F(x, y)$ on $G/K \times G/K$, such that for each x, the function

$$y \mapsto F(x, y) \text{ is in } L^1(\Gamma\backslash G),$$

we have the convolution

$$(F * \psi)(x) = \int\limits_{\Gamma\backslash G} F(x, y)\psi(y)dy = [F_x, \psi]_{\Gamma\backslash G}.$$

THEOREM 3.1. *Let* $\psi \in \mathrm{BC}(\Gamma\backslash G/K)$ *as above. Let*

$$\mathrm{Tr}_{\Gamma_{U_P}\backslash U_P}(\mathbf{K}_{\Gamma\backslash\mathbf{X}})(t, x, y) = \int\limits_{\Gamma_{U_P}\backslash U_P} \mathrm{Tr}_\Gamma(\mathbf{K})(t, ux, y)du.$$

If $\mathrm{Tr}_{\Gamma_{U_P}\backslash U_P}(\mathbf{K}_{\Gamma\backslash\mathbf{X}}) * \psi = 0$, *then* ψ *is* P*-cuspidal.*

PROOF. By assumption,

$$0 = \int\limits_{\Gamma\backslash G} \int\limits_{\Gamma_{U_P}\backslash U_P} \mathrm{Tr}_\Gamma(\mathbf{K})(t, ux, y)\psi(y)dudy.$$

We interchange the order of integration, and let $t \to 0$. Taking the limit under the integral sign is valid because the total integral of the heat kernel on $\Gamma\backslash G$ is 1, and one can apply the dominated convergence theorem. The Dirac property of the heat kernel then concludes the proof.

Next we strive toward Theorem 3.5. We shall use the Eisenstein series with a notation separating out the variables involved, namely

$$\boxed{E_P(\mathrm{Tr}_{\Gamma_{G_P}}(\mathbf{K}_{\mathbf{X}_{G_P}}), \chi_\zeta, \overline{\chi_\zeta})(t, x, y) = E_{P,\Gamma,\mathbf{K}}(\zeta, \bar{\zeta}, t, x, y),}$$

which we convolve with a function $\psi \in \mathrm{BC}(\Gamma\backslash G/K)$, that is,

$$\boxed{(E_P(\mathrm{Tr}_{\Gamma_{G_P}}(\mathbf{K}_{\mathbf{X}_{G_P}}), \zeta, \bar{\zeta}) *_{\Gamma\backslash G} \psi)(t, x) = \int\limits_{\Gamma\backslash G} E_{P,\Gamma,\mathbf{K}}(\zeta, \bar{\zeta}, t, x, y)\psi(y)dy_{\Gamma\backslash G}.}$$

We repeat the pattern of Theorems 1.1, 1.2, 1.3 using the heat kernel instead of F, thus involving two variables (x, y) instead of the single variable x. We start with Mellin-Fourier inversion for $\varphi \in \mathrm{Gauss}(A_P)$. Using (2η) in Chapter 2, §4 we get for $a, b \in A_P, \eta \in \mathfrak{a}^\vee_{P,\mathbf{C}}, \mathrm{Re}(\eta) = \xi$,

$$\varphi(ba^{-1})a^{2\eta} = \int\limits_{\mathfrak{a}^\vee_P} b^{\eta+\mathbf{i}\lambda}a^{\eta-\mathbf{i}\lambda}(\mathbf{M}_{\mathfrak{a}}\varphi_{-\eta})(-\mathbf{i}\lambda)d\lambda. \tag{1}$$

Note that if $\eta = \xi, \zeta = \xi + \mathbf{i}\lambda$, then

$$b^{\xi+\mathbf{i}\lambda}a^{\xi-\mathbf{i}\lambda} = \chi_\zeta(b)\overline{\chi_\zeta(a)}.$$

On the other hand, the Eisenstein series $E_{P,\Gamma,\mathbf{K}}(\chi_{\xi+\mathbf{i}\lambda}, \chi_{\xi-\mathbf{i}\lambda})$ can be extended complex analytically, replacing ξ by the complex variable η (in the domain of absolute convergence of the real part). This will be significant in the second part of the proof of Theorem 3.5. We now run into the two-character Eisenstein series mentioned at the end of Chapter 2, §3. We repeat Chapter 2, Proposition 3.1.

LEMMA 3.2. *For $t \geqq t_0 > 0$, $\mathrm{Tr}_{\Gamma_{G_P}}(\mathbf{K}_{\mathbf{X}_{G_P}})(t, x, y)$ is bounded for x in a compact set and y arbitrary.*

The next lemma corresponds to Theorem 1.2(i). Using (1) above perturbs the Eisenstein series from what the theorem was previously. This perturbation comes from the two $\mathbf{X}$-variables of the heat kernel, and gives rise to an added structure of Dirichlet series.

The integral of the next lemma, with a specially chosen function $\varphi = \varphi_t$, will play an essential role later, starting with §4 below.

LEMMA 3.3. *Let $\xi > 2\rho_P$ and let $\varphi \in \mathrm{Gauss}(A_P)$ (P-**ADM 2**). Then for $x, y \in G$,*

$$\int\limits_{\mathrm{Re}(\zeta)=\xi} E_{P,\Gamma,\mathbf{K}}(\chi_\zeta, \overline{\chi_\zeta})(t,x,y)(\mathbf{M}_{\mathfrak{a}}\varphi)(-\zeta)d\mathrm{Im}(\zeta)$$

$$= \sum_{\gamma_1,\gamma_2 \in \Gamma_P\backslash\Gamma} \mathrm{Tr}_{\Gamma_{G_P}}(\mathbf{K}_{\mathbf{X}_{G_P}})(t, (\gamma_1 x)_{\mathbf{X}_{G_P}}, (\gamma_2 y)_{\mathbf{X}_{G_P}})$$

$$\cdot \varphi((\gamma_1 x)_{A_P}(\gamma_2 y)_{A_P}^{-1})(\gamma_2 y)_{A_P}^{2\xi}.$$

LEMMA 3.3η. *Let $\xi > 2\rho_P$ and let $\varphi \in \mathrm{Gauss}(A_P)$ (P-**ADM 2**). Then for $x, y \in G$ and $\mathrm{Re}(\eta) = \xi$,*

$$\int\limits_{\mathfrak{a}_P^\vee} E_{P,\Gamma,\mathbf{K}}(\chi_{\eta+\mathbf{i}\lambda}, \chi_{\eta-\mathbf{i}\lambda})(t,x,y)(\mathbf{M}_{\mathfrak{a}}\varphi_{-\eta})(-\mathbf{i}\lambda)d\lambda$$

$$= \sum_{\gamma_1,\gamma_2 \in \Gamma_P\backslash\Gamma} \mathrm{Tr}_{\Gamma_{G_P}}(\mathbf{K}_{\mathbf{X}_{G_P}})(t, (\gamma_1 x)_{\mathbf{X}_{G_P}}, (\gamma_2 y)_{\mathbf{X}_{G_P}})$$

$$\cdot \varphi((\gamma_1 x)_{A_P}(\gamma_2 y)_{A_P}^{-1})(\gamma_2 y)_{A_P}^{2\eta}.$$

PROOF. We let

$$b = (\gamma_1 x)_{A_P} \quad \text{and} \quad a = (\gamma_2 y)_{A_P}.$$

We multiply (1) with $\mathrm{Tr}_{\Gamma_{G_P}}(\mathbf{K}_{\mathbf{X}_{G_P}})(t, (\gamma_1 x)_{\mathbf{X}_{G_P}}, (\gamma_2 y)_{\mathbf{X}_{G_P}})$. We then sum over γ_1, γ_2 and the desired formula drops out.

Next the function ψ enters the picture. We want an inversion theorem which refines Theorem 1.2(ii). With the pair of variables of the heat kernel, the scalar product on $\Gamma\backslash G$ is replaced by a convolution. Let x_P, y_P denote variables in G_P. Let $E = E(x_P, y_P)$ be a function of two variables, and $\psi = \psi(y_P)$ a function of one variable. Then by definition, convolution on G_P is given by the integral

$$(E * \psi)(x_P) = \int\limits_{G_P} E(x_P, y_P)\psi(y_P)dy_P.$$

Actually, the functions will be right K_P-invariant, so the variables can be taken in $\mathbf{X}_{G_P} = G_P/K_P$.

We shall consider the convolution operator defined on $\mathrm{BC}(\Gamma\backslash G/K)$ by convolution with the heat Eisenstein series, that is

$$\psi \mapsto E_{P,\Gamma,\mathbf{K}}(\chi_\zeta, \overline{\chi_\zeta}) *_{\Gamma\backslash G} \psi.$$

We are striving toward Theorem 3.5, which says that ψ is in the kernel of this operator (the convolution is 0) if and only if ψ is P-cuspidal. We characterize a function by its effect as a functional on a space of test functions, which we take to be the space of Gauss functions. Thus we deal with the bilinear product occuring in Lemma 3.3,

$$(2) \qquad \int_{\mathfrak{a}_P^\vee} (E_{P,\Gamma,\mathbf{K}}(\chi_{\eta+\mathbf{i}\lambda}, \chi_{\eta-\mathbf{i}\lambda}) *_{\Gamma\backslash G} \psi)(t,x)(\mathbf{M}_{\mathfrak{a}}\varphi_{-\eta})(-\mathbf{i}\lambda)d\lambda.$$

We express this integral as an Eisenstein (generalized Dirichlet) series, whose coefficients are themselves given by an integral in terms of parabolic coordinates, in a manner similar to that of Theorem 1.1(i).

As in Chapter 1, §7, let δ_P be the modular character on P, so $\delta_P(a) = a^{2\rho_P}$. We abbreviate an exponent which will occur systematically, namely we let

$$\beta(\eta) = 2\eta - 2\rho_P.$$

On the whole we try to avoid such abbreviations, but they become useful as compared with double indices, when we want to write $\varphi_{-\beta(\eta)}$ later. We define the **coefficient function** for $x \in G$,

$$(3) \qquad C_{P,\eta}(\mathbf{K}_t, x, \psi, \varphi) =$$

$$\int_{A_P}\int_{G_P} \mathbf{K}_{\mathbf{X}_{G_P}}(t, x_{\mathbf{X}_{G_P}}, g)\mathrm{Tr}_{\Gamma_{U_P}\backslash U_P}(\psi)(x_{A_P}ag)\varphi(a^{-1})a^{\beta(\eta)}dgda.$$

NOTE. The occurrence of $\varphi(a^{-1})$, instead of $\varphi(a)$ as in the Remark following Theorem 1.1, is due to the presence of $\bar{\zeta}$, and the inversion perturbation coming from (1).

The next theorems show that the bilinear product (2) is equal to an Eisenstein series. These theorems correspond to a mixture of Theorem 1.1 and Theorem 1.2(ii). We first state the most important special case with $\eta = \xi$.

THEOREM 3.4. *With* $\zeta = \xi + \mathbf{i}\lambda$, $\xi > 2\rho_P, \varphi \in \mathrm{Gauss}(A_P)$, $\psi \in \mathrm{BC}(\Gamma\backslash G/K)$, *we have*

$$\int_{\mathrm{Re}(\zeta)=\xi} (E_{P,\Gamma,\mathbf{K}}(\chi_\zeta, \bar{\chi}_\zeta) *_{\Gamma\backslash G} \psi)(t,x)(\mathbf{M}_{\mathfrak{a}}\varphi)(-\zeta)d\mathrm{Im}(\zeta)$$

$$= \sum_{\gamma\in\Gamma_P\backslash\Gamma} C_{P,\xi}(\mathbf{K}_t, \gamma x, \psi, \varphi)(\gamma x)_{A_P}^{\beta(\xi)}.$$

The general case with η reads:

THEOREM 3.4η. *Let* $\psi \in \mathrm{BC}(\Gamma\backslash G/K)$. *Suppose that* $\mathrm{Re}(\eta) > 2\rho_P$, *that is,* $\mathrm{Re}(\beta(\eta)) > 2\rho_P$ *to satisfy* P-**ADM 1**. *Then for* $x \in G/K$ *and all* $\varphi \in \mathrm{Gauss}(A_P)$,

$$\int_{\mathfrak{a}_P^\vee} (E_{P,\Gamma,\mathbf{K}}(\chi_{\eta+\mathbf{i}\lambda}, \chi_{\eta-\mathbf{i}\lambda}) *_{\Gamma\backslash G} \psi)(t,x)(\mathbf{M}_{\mathfrak{a}}\varphi_{-\eta})(-\mathbf{i}\lambda)d\lambda$$

$$= \sum_{\gamma \in \Gamma_P \backslash \Gamma} (\gamma x)_{A_P}^{2\eta - 2\rho_P} \cdot$$

$$\cdot \int_{A_P} \int_{G_P} \mathbf{K}_{\mathbf{X}_{G_P}}(t, (\gamma x)_{\mathbf{X}_{G_P}}, g) \mathrm{Tr}_{\Gamma_{U_P} \backslash U_P}(\psi)((\gamma x)_{A_P} a_g) \varphi(a^{-1}) a^{2\eta - 2\rho_P} adgda$$

$$= \sum_{\gamma \in \Gamma_P \backslash \Gamma} C_{P,\eta}(\mathbf{K}_t, \gamma x, \psi, \varphi)(\gamma x)_{A_P}^{\beta(\eta)}.$$

PROOF. We multiply the left side and right side of the equality in Lemma 2.3η by $\psi(y)$ and integrate over $\Gamma \backslash G$. We thus obtain:

$$\int_{\Gamma \backslash G} \int_{\mathfrak{a}_P^\vee} E_{P,\Gamma,\mathbf{K}}(\chi_{\eta + \mathbf{i}\lambda}, \chi_{\eta - \mathbf{i}\lambda})(t, x, y) \psi(y) (\mathbf{M}_{\mathfrak{a}} \varphi_{-\eta})(-\mathbf{i}\lambda) d\lambda dy_{\Gamma \backslash G}$$

$$= \int_{\Gamma \backslash G} \sum_{\gamma_2} \sum_{\gamma_1} \mathrm{Tr}_{\Gamma_{G_P}}(\mathbf{K}_{\mathbf{X}_{G_P}})(t, (\gamma_1 x)_{\mathbf{X}_{G_P}}, (\gamma_2 y)_{\mathbf{X}_{G_P}}) \psi(y) \cdot$$

$$\cdot \varphi((\gamma_1 x)_{A_P} (\gamma_2 y)_{A_P}^{-1}) (\gamma_2 y)_{A_P}^{2\eta} dy_{\Gamma \backslash G}.$$

The sum is taken for $\gamma_1, \gamma_2 \in \Gamma_P \backslash \Gamma$. By §1 (3) this last expression is

$$= \int_{A_P} \int_{G_P} \sum_{\gamma_1} \mathbf{K}_{\mathbf{X}_{G_P}}(t, (\gamma_1 x)_{\mathbf{X}_{G_P}}, g) \mathrm{Tr}_{\Gamma_{U_P} \backslash U_P}(\psi)(ag) \cdot$$

$$\cdot \varphi((\gamma_1 x)_{A_P} a^{-1}) a^{2\eta} \delta_P^{-1}(a) dgda.$$

We make the $(\gamma_1 x)_{A_P}$-translation of a in the A_P-integral to conclude the proof.

REMARK. We note that the sums of Theorem 3.4η are Eisenstein series and can be viewed as generalized Dirichlet series. The coefficients are given by a double integral. If we use $\mathrm{Tr}_{\Gamma_{G_P}}(\mathbf{K}_{\mathbf{X}_{G_P}})$ in the integrand, we change the integral over G_P to an integral over $\Gamma_{G_P} \backslash G_P$. The other terms are left Γ_{G_P}-invariant, so are unaffected by such a trace.

The notion of P-cuspidality was defined in terms of the U_P-variable. The next theorem gives an equivalent definition in terms of the $A_P G_P$-variables. It is a deepening of Theorem 1.3. It does not involve merely a general function F, but it involves the Dirac property of the heat kernel. The heat equation does not yet appear.

THEOREM 3.5. *Let P be a standard reduced parabolic subgroup of G. Let $\psi \in \mathrm{BC}(\Gamma \backslash G/K)$. The following conditions are equivalent.*

P-**CUS 1.** *The function ψ is P-cuspidal, i.e. $\mathrm{Tr}_{\Gamma_{U_P} \backslash U_P}(\psi) = 0$.*

P-**CUS 2.** *There exists $\xi \in \mathfrak{a}_P^\vee$, $\xi > 2\rho_P$ such that for all $\zeta \in \mathfrak{a}_{P,\mathbf{C}}^\vee$ with $\mathrm{Re}(\zeta) = \xi$ we have (with convolution on $\Gamma \backslash G$):*

$$E_P(\mathrm{Tr}_{\Gamma_{G_P}}(\mathbf{K}_{\mathbf{X}_{G_P}}), \chi_\zeta, \overline{\chi_\zeta}) * \psi = 0.$$

P-**CUS 3.** *For all values of* $\xi > 2\rho_P$ *we have for* $\mathrm{Re}(\zeta) = \xi$,

$$E_P(\mathrm{Tr}_{\Gamma_{G_P}}(\mathbf{K}_{\mathbf{X}_{G_P}}), \chi_\zeta, \overline{\chi_\zeta}) * \psi = 0.$$

PROOF. Assume P-**CUS 1.** Then the expression on the right of the equality in Theorem 3.4 is 0, whence so is the expression on the left. For each (x, t) the function $\lambda \mapsto (E_{P,\Gamma,\mathbf{K}}(\chi_{\xi+i\lambda}, \chi_{\xi-i\lambda}) * \psi)(t, x)$ is orthogonal to the Gauss space of $\mathfrak{a}_P^\vee$, and is therefore equal to 0, thus proving one implication, even the strong one namely P-**CUS 3.**

So far we have not used any of the essential properties of the heat kernel, we have used only group invariance properties. We now shall use the Dirac property for the converse implication. We assume P-**CUS 2** and want to prove P-**CUS 1.** Then the sum on the right of the first equality in Theorem 3.4η is 0. *Abbreviate* $\mathrm{Tr}_{\Gamma_{U_P}\backslash U_P}(\psi)$ *by* $\mathrm{Tr}(\psi)$. Formally, let $t \to 0$. By the Dirac property,

$$0 = \sum_{\gamma\in\Gamma_P\backslash\Gamma} (\gamma x)_{A_P}^{\beta(\eta)} \int_{A_P} \mathrm{Tr}(\psi)((\gamma x)_{A_P} a(\gamma x)_{\mathbf{x}_{G_P}}) \varphi_{-\beta(\eta)}(a^{-1}) da$$

for arbitrary $\varphi \in \mathrm{Gauss}(A_P)$, and therefore arbitrary $\varphi_{-\beta(\eta)}$. In particular, we can take $\varphi_{-\beta(\eta)}$ to be the euclidean Gaussian giving rise to the heat kernel, say h_t with $t > 0$, and let $t \to 0$. From the Dirac property, we get

$$0 = \sum_{\gamma\in\Gamma_P\backslash\Gamma} (\gamma x)_{A_P}^{\beta(\eta)} \mathrm{Tr}(\psi)((\gamma x)_{A_P} \mathbf{x}_{G_P}).$$

This relation is formally valid for $\beta(\eta) \in 2\xi - 2\rho_P + \mathbf{i}\mathfrak{a}_P^\vee$. Therefore it is valid in the half space of absolute convergence of the character Eisenstein series

$$\sum_{\gamma\in\Gamma_P\backslash\Gamma} (\gamma x)_{A_P}^{\zeta}, \quad \mathrm{Re}(\zeta) > 2\rho_P.$$

We now want to conclude that the coefficient

$$c_1(x) = \mathrm{Tr}(\psi)(x_{A_P} \mathbf{x}_{G_P})$$

of the above generalized Dirichlet series is equal to 0. For this we need a lemma.

LEMMA 3.6. *There exists a dense set of elements* $x \in G/K$ *such that if*

$$(\gamma x)_{A_P} = x_{A_P} \text{ for some } \gamma \in \Gamma_P\backslash\Gamma$$

then $\gamma \in \Gamma_P$.

PROOF. For each $\gamma \in \Gamma_P\backslash\Gamma$, $\gamma \notin \Gamma_P$, the equation $(\gamma x)_{A_P} = x_{A_P}$ determines a real analytic subset of G/K, whose complement is open dense. The intersection of these complements is therefore dense, as desired. Note that in the most classical case of $\mathrm{SL}_2(\mathbf{Z})$ acting on the upper half plane G/K, the denumerable family of the above equations is $|cz + d|^2 = 1$, which is a family of ordinary circles.

The function $x \mapsto \mathrm{Tr}(\psi)(x_{A_P} \mathbf{x}_{G_P})$ is a continuous function on G. To show it is the 0 function, it suffices to prove that its values on the dense set of Lemma 3.6

are 0, so assume x is in this dense set. Let

$$c_\gamma(x) = \mathrm{Tr}(\psi)((\gamma x)_{A_P}\mathbf{x}_{G_P}).$$

Then we have a relation for a generalized Dirichlet series in several variables

$$0 = c_1(x)x_{A_P}^\zeta + \sum_{\substack{\gamma\in\Gamma_P\backslash\Gamma \\ \gamma\notin\Gamma_P}} c_\gamma(x)(\gamma x)_{A_P}^\zeta \quad \text{for } \mathrm{Re}(\zeta) \text{ sufficiently large.}$$

The series converges absolutely for a given $\mathrm{Re}(\zeta)$ large. Terms with equal values of $(\gamma x)_{A_P}$ can then be combined for $\gamma \notin \Gamma_P$. Lemma 3.8 below applies to conclude the proof of Theorem 3.5.

We shall complement Theorem 3.5 by another orthogonality result with respect to the "constant term" of the Eisenstein series (to be defined) in a subsequent work.

Appendix to §3.3.

We deal here with the justification for the assertion concerning the generalized Dirichlet series made at the end of the proof of Theorem 3.5. This is just a matter of exponential series, or a version of Artin's theorem concerning the linear independence of characters in a series setting. Similar considerations have occurred as in lemmas of Harish-Chandra reproduced in [**JoL 01a**], Chapter VIII, §0. Here we need only the simplest considerations, which we present ab ovo.

We start with series in one complex variable, just to see what's going on.

LEMMA 3.7. *Let* $\{c_j\}$ $(j = 1, 2, \dots,)$ *be a sequence of complex numbers. Let* $\{y_j\}$ *be a sequence of distinct real numbers. Suppose there is some* σ_0 *such that*

$$0 = \sum_{j=1}^{\infty} c_j e^{sy_j}$$

for all complex s *with* $\mathrm{Re}(s) = \sigma_0$*, the series being assumed to converge absolutely. Then* $c_j = 0$ *for all* j.

PROOF. Say $c_1 \neq 0$. We multiply the series by $(c_1 e^{sy_1})^{-1}$, so that without loss of generality, we may assume $c_1 = 1$ and $y_1 = 0$. So the series looks like

$$0 = 1 + \sum_{j=2}^{\infty} c_j e^{sy_j}.$$

Write $s = \sigma_0 + \mathbf{i}t$. Let $c_j' = c_j e^{\sigma_0 y_j}$. We can rewrite the sum in the form

$$0 = 1 + \sum_{j=2}^{\infty} c_j' e^{\mathbf{i}ty_j}.$$

Given ε, there exists N such that $\sum_{j=N+1}^{\infty} |c'_j| < \varepsilon$, and so for all t,

$$\left| \sum_{j=1}^{N} c'_j e^{\mathbf{i}ty_j} \right| < \varepsilon.$$

Let $f(t)$ be the sum on the left. A direct computation shows that

$$\sum_{j=1}^{N} |c'_j|^2 = \lim_{T\to\infty} \frac{1}{T} \int_0^T |f(t)|^2 dt \leqq \varepsilon^2.$$

Since $c'_1 = 1$, this is a contradiction which proves the lemma.

The pattern in the higher dimensional case will be exactly the same, with slight notational additions since we deal with r coordinates.

LEMMA 3.8. *Let $\mathfrak{a}$ be a finite dimensional real vector space of dimension r. Let $\{c_j\}$ be a sequence of complex numbers, and $\{H_j\}$ a sequence of distinct elements of $\mathfrak{a}$. Let $\xi \in \mathfrak{a}^\vee$. Suppose that for all $\zeta \in \mathfrak{a}^\vee_{\mathbf{C}}$ with* $\mathrm{Re}(\zeta) = \xi$ *we have*

$$0 = \sum_{j=1}^{\infty} c_j e^{\zeta(H_j)},$$

the series being assumed to converge absolutely. Then $c_j = 0$ for all j.

PROOF. Suppose $c_1 \neq 0$. We multiply the series by $(c_1 e^{\zeta(H_j)})^{-1}$, so that without loss of generality, we may assume $c_1 = 1$ and $H_1 = 0$. So the series looks like

$$0 = 1 + \sum_{j=2}^{\infty} c_j e^{\zeta(H_j)}.$$

Let $\{\lambda_1, \dots, \lambda_r\}$ be a basis for $\mathfrak{a}^\vee$. Write

$$\zeta = \xi + \mathbf{i}\lambda \quad \text{with} \quad \lambda = t_1\lambda_1 + \ldots + t_r\lambda_r,$$

with coefficients t_ν being real. Let $c'_j = c_j e^{\xi(H_j)}$. We have

$$0 = \sum_{j=1}^{\infty} c'_j e^{\mathbf{i}(t_1\lambda_1(H_j) + \ldots + t_r\lambda_r(H_j))}.$$

Given ε, there exists N such that

$$\left| \sum_{j=1}^{N} c'_j e^{\mathbf{i}\lambda(H_j)} \right| < \varepsilon.$$

Let $f(t_1, \dots, t_r) = \sum_{j=1}^{N} c'_j e^{\mathbf{i}\lambda(H_j)}$, so $|f(t_1, \dots, t_r)| < \varepsilon$. Then

$$|f(t_1, \dots, t_r)|^2 = \sum_{j,k=1}^{N} c'_j \overline{c'_k} e^{\mathbf{i}\lambda(H_j - H_k)} < \varepsilon^2.$$

We now claim that

$$\sum_{j=1}^{N} |c'_j|^2 = \lim_{T\to\infty} \frac{1}{T}\int_0^T \cdots \frac{1}{T}\int_0^T |f(t_1,\ldots,t_r)|^2 dt_1 \ldots dt_r < \varepsilon^2.$$

We have to show that the cross terms involving $c'_j\overline{c'_k}$ with $j \neq k$ tend to 0 when $T \to \infty$. This is where we use the hypothesis that $H_1, \ldots, H_N$ are distinct. Given $j \neq k$, there is an index $\nu = 1, \ldots, r$ such that $\lambda_\nu(H_j - H_k) \neq 0$. Let $y_{jk\nu} = \lambda_\nu(H_j - H_k)$. Then

$$\frac{1}{T}\int_0^T e^{\mathbf{i}t_\nu \lambda_\nu(H_j-H_k)}dt = \frac{1}{T}\frac{e^{\mathbf{i}Ty_{jk\nu}}}{\mathbf{i}y_{jk\nu}}.$$

Hence the cross terms with $j \neq k$ disappear as $T \to \infty$, and what remains is the sum

$$\sum_{j=1}^{N} |c'_j|^2,$$

thus proving the claim. Since one of the terms in the sum is 1, we get a contradiction, which proves Lemma 3.8.

3.4. The family of anticuspidal operators $J_{P,\Gamma,\xi,t}$

One partial goal is to develop one-parameter semigroups associated with parabolic subgroups. Here we merely record some initial conditions. Recall the notation for $\zeta \in \mathfrak{a}^\vee_{P,\mathbf{C}}$, $\mathrm{Re}(\zeta) > 2\rho_P$,

$$E_P(\mathrm{Tr}_{\Gamma_{G_P}}(\mathbf{K}_{\mathbf{X}_{G_P}}), \chi_\zeta, \overline{\chi_\zeta})(t,x,y) = E_{P,\Gamma,\mathbf{K}}(\zeta,\bar{\zeta},t,x,y).$$

Note that $E_{P,\Gamma,\mathbf{K}}$ denotes the heat Eisenstein series as a function of all its variables ζ, t, x, y. The subscript $\mathbf{K}$ indicates twisting with the heat kernel. We may view convolution as a scalar product, integration being taken with respect to one of the variables. In the present case, it is the last variable y. Thus we may write in two ways,

$$\int_{\Gamma\backslash G} E_P(\mathrm{Tr}_{\Gamma_{G_P}}(\mathbf{K}_{\mathbf{X}_{G_P}}), \chi_\zeta, \overline{\chi_\zeta})(t,x,y)\psi(y)dy$$
$$= (E_{P,\Gamma,\mathbf{K}} * \psi)(\zeta,\bar{\zeta},t,x) = [E_{P,\Gamma,\mathbf{K}}(\zeta,\bar{\zeta},t,x),\psi]_{\Gamma\backslash G}.$$

The convolution is on $\Gamma\backslash G$, i.e. $E_{P,\Gamma,\mathbf{K}} *_{\Gamma\backslash G} \psi$. Theorem 3.4 tells us that

$$\text{(1a)}\qquad \int_{\mathrm{Re}(\zeta)=\xi} (E_{P,\Gamma,\mathbf{K}} * \psi)(\zeta,\bar{\zeta},t,x)(\mathbf{M}_\mathfrak{a}\varphi)(-\zeta)(d\mathrm{Im}(\zeta)$$
$$= \sum_{\gamma\in\Gamma_P\backslash\Gamma} (\gamma x)_{A_P}^{2\xi-2\rho_P} C_{P,\xi}(t,\gamma x,\psi,\varphi).$$

where

$$\text{(1b)}\qquad C_{P,\xi}(t,z,\psi,\varphi)$$

$$= \int_{A_P} \int_{\Gamma_{G_P}\backslash G_P} \mathrm{Tr}_{\Gamma_{G_P}}(\mathbf{K}_{\mathbf{X}_{G_P}})(t, z_{\mathbf{X}_{G_P}}, g)\mathrm{Tr}_{\Gamma_{U_P}\backslash U_P}(\psi)(z_{A_P}ag)\varphi(a^{-1})a^{2\xi-2\rho_P}dgda.$$

We may apply this formula to the case when $\mathbf{M}_{\mathfrak{a}}\varphi$ has a special shape, as in Chapter 2, Proposition 4.3. Thus we are *given* an element $\mu \in \mathfrak{a}_P^\vee$. Let φ_t be the Gauss function such that

$$\mathbf{M}_{\mathfrak{a}}\varphi_t(-\zeta) = e^{(\langle\zeta,\zeta\rangle - \langle\zeta,\mu\rangle)t} = e^{q^\vee(-\zeta)t},$$

where $q^\vee(\zeta) = \langle\zeta,\zeta\rangle + \langle\zeta,\mu\rangle$. In Chapter 4, §2 and §3, we shall take $\mu = \tau_P$ (the $(\mathfrak{a}_P, \mathfrak{n}_P)$-trace), and then

$$q^\vee(-\zeta) = \langle\zeta,\zeta\rangle - \langle\zeta,\tau_P\rangle$$

is the eigenvalue of the Casimir operator on the character χ_ζ.

We specialize the integral of Lemma 3.3 by using the special function φ_t, and **define**

$$\boxed{J_{P,\Gamma,\xi,t} = E_{P,\Gamma,\mathbf{K},t} *_\xi (\mathbf{M}_{\mathfrak{a}}\varphi_t)^-},$$

which we write down more fully, namely for $t > 0$,

(2a) $$\boxed{J_{P,\Gamma,\xi,t}(x,y) = \int_{\mathrm{Re}(\zeta)=\xi} E_{P,\Gamma,\mathbf{K},t}(\zeta,\bar{\zeta},x,y)(\mathbf{M}_{\mathfrak{a}}\varphi_t)(-\zeta)d\mathrm{Im}(\zeta)}.$$

The effect on a function $\psi \in \mathrm{BC}(\Gamma\backslash G/K)$ is the integral of Theorem 3.4, namely

(2b) $$J_{P,\Gamma,\xi,t}(\psi)(x) = \int_{\mathrm{Re}(\zeta)=\xi} (E_{P,\Gamma,\mathbf{K},t} * \psi)(\zeta,\bar{\zeta},x)(\mathbf{M}_{\mathfrak{a}}\varphi_t)(-\zeta)d\mathrm{Im}(\zeta).$$

Of course, μ should also be included in the notation, but we suppose μ fixed. As mentioned above, in Chapter 4 and subsequently we take $\mu = \tau_P = 2\rho_P$.

Directly from the definitions, formula (2b) can be rewritten in the form

(3) $$\begin{aligned} J_{P,\Gamma,\xi,t}(\psi)(x) &= ((M_{\mathfrak{a}}\varphi_t)^- *_\xi E_{P,\Gamma,\mathbf{K},t} * \psi)(x) \\ &= \text{integral expression (1a).} \end{aligned}$$

Next, consider (1b). By Chapter 2, Proposition 4.3,

(4) $$\varphi_t(a^{-1}) = (2t)^{-r/2}e^{-(\log a)^2/4t}a^{-\mu/2}e^{-\mu^2 t/4},$$

where we abbreviate $\mu^2 = \langle\mu,\mu\rangle$ and $H^2 = \langle H, H\rangle$. Let $dH, da = d^*a$ be Haar measures for which Fourier inversion holds, normalized as usual in a euclidean space (ordinary Lebesgue measure with respect to an orthonormal basis times the factor $(2\pi)^{-r/2}$). Then the heat Gaussian and the heat kernel on A_P are respectively

(5a) $$h_t(a) = (2t)^{-r/2}e^{-(\log a)^2/4t}$$

(5b) $$h_t(ab^{-1}) = h_t(a,b) = (2t)^{-r/2}e^{-(\log ab^{-1})^2/4t}.$$

Thus $h_t(a^{-1}) = h_t(a)$ and

(5c) $$\varphi_t(a^{-1}) = h_t(a)a^{-\mu/2}e^{-\mu^2 t/4}.$$

We can then rewrite (1b) in a way which exhibits the role of the heat kernel on the two factors $\Gamma_{G_P}\backslash G_P$ and A_P. The double integral of (1b) leads us to consider the product manifold

$$S_P = A_P \times (\Gamma_{G_P}\backslash G_P).$$

We deal with right K_{G_P}-invariant functions on this space. The heat kernel on a product is the tensor product of the heat kernels on the factors. Let $\mathbf{K}_{S_P}$ be the heat kernel on S_P for right K_{G_P}-invariant functions. For $z \in G/K, g \in G_P$, and $a, b \in A_P$, we then have

$$\mathbf{K}_{S_P}(t, bz\mathbf{x}_{G_P}, ag) = \mathrm{Tr}_{\Gamma_{G_P}}(\mathbf{K}_{\mathbf{X}_{G_P}})(t, z\mathbf{x}_{G_P}, g)h_t(a, b). \tag{6}$$

Note that $\mathrm{Tr}_{\Gamma_{U_P}\backslash U_P}(\psi)$ is a function on S_P. Let $f_{\xi,z}$ be the right K_{G_P}-invariant function on S_P given by

$$f_{\xi,z}(ag) = \mathrm{Tr}_{\Gamma_{U_P}\backslash U_P}(\psi)(z_{A_P}ag)a^{2\xi-2\rho_P+\mu/2}. \tag{7}$$

Then from the definitions and the integral expression (1b) we see directly that

$$C_{P,\xi}(t, z, \psi, \varphi_t) = (\mathbf{K}_{S_P} * f_{\xi,z})(t, e_{A_P}a\mathbf{x}_{G_P})e^{-\mu^2 t/4}. \tag{8}$$

Therefore (1a) and (1b) now take on the more structural aspect of the next theorem.

THEOREM 4.1. *Let $\psi \in \mathrm{BC}(\Gamma\backslash G/K)$ and let $f_{\xi,z}$ be the function defined in (7). Then*

$$J_{P,\Gamma,\xi,t}(\psi)(x) = \sum_{\gamma\in\Gamma_P\backslash P} (\gamma x)_{A_P}^{2\xi-2\rho_P}(\mathbf{K}_{S_P} * f_{\xi,\gamma x})(t, e_{A_P}(\gamma x)\mathbf{x}_{G_P})e^{-\mu^2 t/4}.$$

The expression of Theorem 4.1 is in a form where we can determine the initial conditions. We define

$$J_{P,\Gamma,\xi,0}(\psi)(x) = \sum_{\gamma\in\Gamma_P\backslash\Gamma} (\gamma x)_{A_P}^{2\xi-2\rho_P}\mathrm{Tr}_{\Gamma_{U_P}\backslash U_P}(\psi)((\gamma x)_{A_P}\mathbf{x}_{G_P}) \tag{9}$$

or

$$J_{P,\Gamma,\xi,0}(\psi) = E_P(\mathrm{Tr}_{\Gamma_{U_P}\backslash U_P}(\psi), \chi_{2\xi-2\rho_P}).$$

REMARK. The above notation is convenient, and extends the notation of Chapter 2, §2, where we consider F-Eisenstein series with functions F on G_P. Here we have the extra factor A_P which has to be taken into account, so the projection of γx in each term is on $A_P\mathbf{X}_{G_P}$.

THEOREM 4.2. *Let $\psi \in \mathrm{BC}(\Gamma\backslash G/K)$ and $\xi > 2\rho_P$. Then*

$$\lim_{t\to 0} J_{P,\Gamma,\xi,t}(\psi)(x) = J_{P,\Gamma,\xi,0}(\psi)(x) \text{ for all } x \in G.$$

PROOF. The Dirac property of the heat kernel shows from Theorem 4.1 that the limit is precisely the generalized Dirichlet series $J_{P,\Gamma,\xi,0}(\psi)$. The A_P-component of the heat kernel on S_P is in the Gauss space, and the usual Dirac property applies when we convolve the heat kernel with a character, which has only exponential linear growth. Since $\chi(e_A) = 1$, we see that the term arising from the character disappears in the limit $t \to 0$. So does the term $e^{-\mu^2 t/4}$. This concludes the proof of Theorem 4.2.

REMARK. The initial condition is independent of μ.

CHAPTER 4

Applications of the Heat Equation

In this chapter we determine the direct image of the Casimir operator on parabolics, and use this to show how the heat Eisenstein series satisfy the heat equation or are eigenfunctions of Casimir, depending on normalizations. We then show how the uniqueness of solutions of the heat equation leads to the deeper study of the operator $J_{P,\Gamma,\xi,t}$ defined in Chapter 3, §4, and to the need for the analytic continuation of the heat Eisenstein series in the variable ζ.

4.1. Parabolics and the $(\mathfrak{a}, \mathfrak{n})$-characters

The torus group T_P will now play a role. Recall that T_P consists of the diagonal matrices constant in each block, with components of absolute value 1, and of course having determinant 1. For the Lie algebra, we have

$$\mathfrak{t}_P = \mathrm{Lie}(T_P) = \mathbf{i}\mathfrak{a}_P.$$

As usual, let $\mathfrak{g}_\alpha$ be the α-eigenspace in $\mathfrak{g} = \mathrm{Lie}(G), G = \mathrm{SL}_n(\mathbf{C})$, for the Lie action of $\mathfrak{a}$ (bracket action). The 0-eigenspace is $\mathfrak{a} + \mathbf{i}\mathfrak{a}$, and one usually reserves α for the eigencharacters which occur in the semisimple decomposition of $\mathfrak{n}$. The set of these characters is denoted by $\mathcal{R}(\mathfrak{n})$. A basis of $\mathfrak{n}_\alpha$ in the real case is given by the matrix $E_\alpha (= E_{ij}, i < j)$. In the complex case, an $\mathbf{R}$-basis is given by $E_\alpha, \subset E_\alpha$. We recall the commutation rule

$$[E_\alpha, E_{-\alpha}] = H_\alpha \quad \text{for } \alpha \in \mathcal{R}(\mathfrak{n}).$$

If $E_\alpha = E_{ij}$, then $H_\alpha = E_{ii} - E_{jj}$.

Using the notation of Chapter 1, §5, for a reduced standard parabolic P,

$$\mathfrak{g}_{G_P} = (\mathfrak{a}_{G_P} + \mathbf{i}\mathfrak{a}_{G_P}) + \sum_{\alpha \in \mathcal{R}(\mathfrak{n}_{G_P})} (\mathfrak{g}_\alpha + \mathfrak{g}_{-\alpha}) \tag{1}$$

is an orthogonal direct sum decomposition. For each block, this is the standard decomposition of diagonal matrices and the pair of upper and lower diagonal matrices. Note that $\mathfrak{a}_{G_P} + \mathbf{i}\mathfrak{a}_{G_P}$ is the 0-eigenspace for the regular action of $\mathfrak{a}_{G_P}$ on $\mathfrak{g}_{G_P}$.

As usual, we use the real trace form B on $\mathfrak{g}$, so

$$B(Z, W) = \mathrm{Re}\ \mathrm{tr}(ZW).$$

So on $\mathfrak{a}, B$ is the trace form.

We let $\mathcal{B}$ denote a basis. So we let $\mathcal{B}(\mathfrak{a}_P)$ be a B-orthonormal basis of $\mathfrak{a}_P$ for the trace form. We can of course complete $\mathcal{B}(\mathfrak{a}_P)$ with an orthonormal basis $\mathcal{B}(\mathfrak{a}_{G_P})$ according to the decomposition of Chapter 1, Lemma 4.3. Thus

$$\mathfrak{a} = \mathfrak{a}_P + \mathfrak{a}_{G_P}$$

is an orthogonal direct sum, with basis equal to the union $\mathcal{B}(\mathfrak{a}_P) \cup \mathcal{B}(\mathfrak{a}_{G_P})$. We let

$$\mathcal{B}(\mathfrak{a}_P, \mathfrak{t}_P) = \mathcal{B}(\mathfrak{a}_P) \cup \mathcal{B}(\mathfrak{t}_P),$$

and use similar notation for the other direct sums.

For $G = \mathrm{SL}_n(\mathbf{R})$, a basis of $\mathfrak{g}_G$ is given by

(2**R**) $$\mathcal{B}(\mathfrak{g}_{G_P}) = \mathcal{B}(\mathfrak{a}_{G_P}) \cup \{E_\alpha, E_{-\alpha}\} \quad \text{with } \alpha \in \mathcal{R}(\mathfrak{n}_{G_P}).$$

For $G = \mathrm{SL}_n(\mathbf{C})$, a basis for $\mathfrak{g}_{G_P}$ is given by

(2**C**) $$\mathcal{B}(\mathfrak{g}_{G_P}) = \mathcal{B}(\mathfrak{a}_{G_P}, \mathfrak{t}_{G_P}) \cup \{E_\alpha, E_{-\alpha}, \mathbf{i}E_\alpha, \mathbf{i}E_{-\alpha}\}$$
$$\text{with } \alpha \in \mathcal{R}(\mathfrak{n}_{G_P}).$$

We have an orthogonal direct sum decomposition

(3) $$\mathfrak{g} = (\mathfrak{n}_P + {}^t\mathfrak{n}_P) \perp (\mathfrak{a}_P + \mathbf{i}\mathfrak{a}_P) \perp \mathfrak{g}_{G_P}.$$

However, within the parentheses, $\mathfrak{n}_P$ is not orthogonal to ${}^t\mathfrak{n}_P$, but $\mathfrak{a}_P$ is orthogonal to $\mathbf{i}\mathfrak{a}_P$. Note that $\mathfrak{a}_P + \mathbf{i}\mathfrak{a}_P$ consists of scalar diagonal matrices in each block, so their orthogonality with elements of $\mathfrak{g}_{G_P}$ comes from the definition that these elements have trace 0 in each block.

In Chapter 1, §5 we already noted the disjoint union

$$\mathcal{R}(\mathfrak{n}) = \mathcal{R}(\mathfrak{n}_{G_P}) \cup \mathcal{R}(\mathfrak{n}_P),$$

and the eigenspace decomposition

(4) $$\mathfrak{n}_P = \sum_{\alpha \in \mathcal{R}(\mathfrak{n}_P)} \mathfrak{g}_\alpha.$$

From the full Iwasawa decomposition, we recall the element

(5) $$\rho_G = \frac{1}{2} \sum_{\alpha \in \mathcal{R}(\mathfrak{n}_G)} m(\alpha)\alpha = \frac{1}{2}\tau_G.$$

Similarly, we have the elements

(5P) $$\rho_{G_P} = \frac{1}{2} \sum_{\alpha \in \mathcal{R}(\mathfrak{n}_{G_P})} m(\alpha)\alpha \quad \text{and} \quad \rho_P = \frac{1}{2} \sum_{\alpha \in \mathcal{R}(\mathfrak{n}_P)} m(\alpha)\alpha$$

Recall that $m(\alpha) = 2$ for $G = \mathrm{SL}_n(\mathbf{C})$, and $= 1$ for $G = \mathrm{SL}_n(\mathbf{R})$.

We had the orthogonal decomposition

$$\rho_G = \rho_{G_P} + \rho_P,$$

and ρ_P is $\mathcal{S}(\mathfrak{n}_P)$-positive. See Lemmas 5.1 and 5.2 in Chapter 1.

From now on, we pick the functional $\mu = \tau_P$ to apply the computations of Chapter 2, §7, where we stated that μ would eventually be selected canonically with respect to P. Using the duality symbol, we may write by definition

$$\tau_P = 2\rho_P = H_{\tau_P}^\vee.$$

We follow previous notation, whereby for $\lambda \in \mathfrak{a}^\vee$, H_λ is the vector (element in $\mathfrak{a}$) representing the functional λ vis a vis the scalar product induced by the trace form on $\mathfrak{a}$; in other words, for all $H \in \mathfrak{a}$,

$$B(H_\lambda, H) = \lambda(H).$$

WARNING. On the complexification $\mathfrak{a}_{\mathbf{C}} = \mathfrak{a} + \mathbf{i}\mathfrak{a}$ there are two natural forms. One of them is the real trace form B, and the other is the $\mathbf{C}$-bilinear extension of $B_{\mathfrak{a}}$ to $\mathfrak{a} + \mathbf{i}\mathfrak{a}$ and the corresponding dual extension $B_{\mathbf{C}}^\vee$ to $\mathfrak{a}^\vee + \mathbf{i}\mathfrak{a}^\vee$. We abbreviate $B_{\mathbf{C}}^\vee(\zeta, \eta) = \langle \zeta, \eta \rangle$. For $H \in \mathfrak{a}$, the differential operator associated to H is denoted by $\tilde{H}$ or $\mathcal{D}(H)$ (derivative in the direction of H). By definition, for a function f on A, we have

$$(\tilde{H}f)(a) = \frac{d}{dt} f(a \cdot \exp(tH)) \Big|_{t=0}.$$

Then by freshman calculus, for $\zeta \in \mathfrak{a}_{\mathbf{C}}^\vee$ and $f(a) = \chi_\zeta(a) = a^\zeta = e^{\zeta(\log a)}$, we get

$$\tilde{H}\chi_\zeta = \zeta(H)\chi_\zeta,$$

so χ_ζ is an eigenfunction of $\tilde{H}$, with eigenvalue $\zeta(H)$. Hence trivially,

$$\tilde{H}^2\chi_\zeta = \zeta(H)^2\chi_\zeta.$$

The next section goes more systematically into these differential operators, on G.

4.2. Direct image of Casimir on parabolics

In this section, we compute the decomposition of Casimir corresponding to the parabolic decomposition of $\mathfrak{g}$. If $Z \in \mathfrak{g}$ we let $\tilde{Z} = \mathcal{D}(Z)$ be the associated left invariant differential operator on G. Here we view G as a real Lie group. For any C^∞ function f on G, we have for $g \in G$,

$$(\tilde{Z}f)(g) = \frac{d}{dt} f(g \exp(tZ)) \Big|_{t=0}.$$

The algebra of operators generated by such $\tilde{Z}$ is called the algebra of **invariant differential operators** (left invariant, that is). One has to be careful to take into account that the association $Z \mapsto \tilde{Z} = \mathcal{D}(Z)$ is not $\mathbf{C}$-linear, only $\mathbf{R}$-linear. Aside from this, the exposition of [**JoL 01**], Chapter II is valid for $\mathrm{SL}_n(\mathbf{C})$ as well as $\mathrm{SL}_n(\mathbf{R})$. We shall use the fact that the association of the differential operator to a vector is a homomorphism of Lie algebras, that is

$$[\tilde{Z}, \tilde{Z}'] = [Z, Z'] \quad \text{for } Z, Z' \in \mathfrak{g}.$$

We now proceed as in [**JoL 01a**], Chapter VII.

For any Lie group G with Lie algebra $\mathfrak{g}$, and a (G)-invariant non-degenerate symmetric bilinear form B on $\mathfrak{g}$, one defines the **Casimir operator** $\omega = \omega_G$ as follows. Let $\{Z_1, \ldots, Z_N\}$ be a basis of $\mathfrak{g}$. Let $\{Z'_1, \ldots, Z'_N\}$ be the dual basis with respect to B, that is $B(Z_i, Z'_j) = \delta_{ij}$. The Casimir operator is defined to be

$$\omega_G = \sum_{i=1}^{N} \tilde{Z}_i \tilde{Z}'_i.$$

This expression is independent of the choice of basis, because the element

$$\sum Z_i \otimes Z'_i$$

in the tensor algebra corresponds to the identity in the natural linear isomorphism $\mathfrak{g} \otimes \mathfrak{g} \to \mathrm{End}(\mathfrak{g})$, so is independent of the choice of basis. Cf. [**JoL 01a**], Chapter VII, §2. In the present case, we are using the real trace form for B. For example, let $\{H_j\}$ $(j = 1, \ldots, n-1)$ be an orthonormal basis of $\mathfrak{a}$. Then the Casimir operator on A is

$$\omega_A = \sum_j \tilde{H}_j^2.$$

We express the Casimir operator on $G = \mathrm{SL}_n(\mathbf{C})$ by using the orthogonal decomposition (3) and (2C) of §1. With the orthonormal basis $\{H_j\}$ as above, a basis of $\mathfrak{g}$ is given by

$$\{H_j, \mathbf{i}H_j\}, \{E_\alpha, E_{-\alpha}, E_\alpha^{(\mathbf{i})}, E_{-\alpha}^{(\mathbf{i})}\}_{\alpha \in \mathcal{R}(\mathfrak{n})} \quad \text{where} \quad E_\alpha^{(\mathbf{i})} = \mathbf{i}E_\alpha.$$

The dual basis with respect to the real trace form is

$$\{H_j, -\mathbf{i}H_j\}, \{E_{-\alpha}, E_\alpha, -E_{-\alpha}^{(\mathbf{i})}, -E_\alpha^{(\mathbf{i})}\}_{\alpha \in \mathcal{R}(\mathfrak{n})}.$$

Hence the **Casimir operator** is

$$\begin{aligned} (1) \qquad \omega &= \omega_{\mathfrak{a}} + \omega_{\mathfrak{t}} + \sum_{\alpha \in \mathcal{R}(\mathfrak{n})} (\tilde{E}_\alpha \tilde{E}_{-\alpha} + \tilde{E}_{-\alpha} \tilde{E}_\alpha) \\ &\quad - \sum_{\alpha \in \mathcal{R}(\mathfrak{n})} (\tilde{E}_\alpha^{(\mathbf{i})} \tilde{E}_{-\alpha}^{(\mathbf{i})} + \tilde{E}_{-\alpha}^{(\mathbf{i})} \tilde{E}_\alpha^{(\mathbf{i})}) \\ &= \omega_{\mathfrak{a}} + \omega_{\mathfrak{t}} + \omega_{\mathrm{unip}}, \end{aligned}$$

where ω_{unip} will be called the **unipotent part** of ω. The two sums are taken for $\alpha \in \mathcal{R}(\mathfrak{n})$.

Instead of the minimal parabolic decomposition we use the *general parabolic decomposition of* §1 (2**C**). We use what's called the **Cartan involution** θ, defined on the Lie algebra by

$$\theta Z = -{}^t\bar{Z}.$$

Then $\mathfrak{k}$ consists of those $Z \in \mathfrak{g}$ such that $\theta Z = Z$ (fixed point set of θ). We let the P-unipotent part of Casimir be

$$\begin{aligned} \omega_{P\mathrm{unip}} &= \omega_{\mathfrak{n}_P + \theta\mathfrak{n}_P} \\ (2) \qquad &= \sum_{\alpha \in \mathcal{R}(\mathfrak{n}_P)} (\tilde{E}_\alpha \tilde{E}_{-\alpha} + \tilde{E}_{-\alpha} \tilde{E}_\alpha) - \sum_{\alpha \in \mathcal{R}(\mathfrak{n}_P)} (\tilde{E}_\alpha^{(\mathbf{i})} \tilde{E}_{-\alpha}^{(\mathbf{i})} + \tilde{E}_{-\alpha}^{(\mathbf{i})} \tilde{E}^{(\mathbf{i})}). \end{aligned}$$

The sums are like those of (1) except that they are taken over $\alpha \in \mathcal{R}(\mathfrak{n}_P)$.

THEOREM 2.1. *Let $G = \mathrm{SL}_n(\mathbf{C})$. With respect to a given reduced standard parabolic subgroup P, the Casimir operator on G has what we call the P-decomposition*

$$\omega_G = \omega_{P\mathrm{unip}} + \omega_{\mathfrak{a}_P + \mathfrak{t}_P} + \omega_{G_P}.$$

PROOF. This is immediate from formula (3) in §1, and the definition of Casimir as recalled above.

We now go into an analysis of each term of the sums occurring in (1) and (2), following a standard pattern. We let $\mathfrak{k} = \mathrm{Lie}(K)$. Cf. [**JoL 01a**], Chapter VII, §3.

LEMMA 2.2. *For $\alpha \in \mathcal{R}(\mathfrak{n})$,*

$$\tilde{E}_\alpha \tilde{E}_{-\alpha} + \tilde{E}_{-\alpha}\tilde{E}_\alpha \equiv 2\tilde{E}_\alpha^2 - \tilde{H}_\alpha \mod \tilde{E}_\alpha \tilde{\mathfrak{k}}.$$

PROOF. We reproduce here a standard computation. We have

$$E_{-\alpha} = {}^t E_\alpha = ({}^t E_\alpha - E_\alpha) + E_\alpha,$$

and $\widetilde{{}^t E_\alpha} - \tilde{E}_\alpha \in \tilde{\mathfrak{k}}$. Hence

$$\tilde{E}_\alpha \tilde{E}_{-\alpha} \equiv \tilde{E}_\alpha^2 \mod \tilde{E}_\alpha \tilde{\mathfrak{k}}.$$

On the other hand,

$$\begin{aligned} \tilde{E}_{-\alpha}\tilde{E}_\alpha = (\tilde{E}_{-\alpha}\tilde{E}_\alpha - \tilde{E}_\alpha \tilde{E}_{-\alpha}) + \tilde{E}_\alpha \tilde{E}_{-\alpha} &= [\tilde{E}_{-\alpha}, \tilde{E}_\alpha] + \tilde{E}_\alpha \tilde{E}_{-\alpha} \\ &= -\tilde{H}_\alpha + \tilde{E}_\alpha \tilde{E}_{-\alpha}. \end{aligned}$$

This proves the lemma.

LEMMA 2.3. *For $\alpha \in \mathcal{R}(\mathfrak{n})$,*

$$\widetilde{E_\alpha^{(\mathrm{i})}}\,\widetilde{E_{-\alpha}^{(\mathrm{i})}} + \widetilde{E_{-\alpha}^{(\mathrm{i})}}\,\widetilde{E_\alpha^{(\mathrm{i})}} \equiv -2(\widetilde{E_\alpha^{(\mathrm{i})}})^2 + \tilde{H}_\alpha \mod \widetilde{E_\alpha^{(\mathrm{i})}}\,\tilde{\mathfrak{k}}.$$

PROOF. We write

$$\widetilde{E_{-\alpha}^{(\mathrm{i})}} = \theta(E_\alpha^{(\mathrm{i})}) = (\theta E_\alpha^{(\mathrm{i})} + E_\alpha^{(\mathrm{i})}) - E_\alpha^{(\mathrm{i})},$$

so

$$\widetilde{E_\alpha^{(\mathrm{i})}}\,\widetilde{E_{-\alpha}^{(\mathrm{i})}} \equiv -(\widetilde{E_\alpha^{(\mathrm{i})}})^2 \mod \widetilde{E_\alpha^{(\mathrm{i})}}\,\tilde{\mathfrak{k}}.$$

On the other hand,

$$\begin{aligned} \widetilde{E_{-\alpha}^{(\mathrm{i})}}\,\widetilde{E_\alpha^{(\mathrm{i})}} &= [\widetilde{E_{-\alpha}^{(\mathrm{i})}}, \widetilde{E_\alpha^{(\mathrm{i})}}] + \widetilde{E_\alpha^{(\mathrm{i})}}\,\widetilde{E_{-\alpha}^{(\mathrm{i})}} \\ &= [E_{-\alpha}^{(\mathrm{i})}, E_\alpha^{(\mathrm{i})}]^\sim - (\widetilde{E_\alpha^{(\mathrm{i})}})^2 \mod \widetilde{E_\alpha^{(\mathrm{i})}}\,\tilde{\mathfrak{k}}. \end{aligned}$$

The first term on the right is $-[E_{-\alpha}, E_\alpha]^\sim = \tilde{H}_\alpha$, which concludes the proof.

The two lemmas and (2) then yield:

PROPOSITION 2.4. *With respect to a parabolic P, we have on* $\mathrm{SL}_n(\mathbf{C})$,

$$\omega_{P\mathrm{unip}} \equiv 2 \sum_{\alpha \in \mathcal{R}(\mathfrak{n}_P)} (\tilde{E}_\alpha^2 + (\widetilde{E_\alpha^{(\mathrm{i})}})^2 - \tilde{H}_\alpha) \mod (\tilde{\mathfrak{n}}_P + \widetilde{\theta \mathfrak{n}_P})\tilde{\mathfrak{k}}.$$

REMARK. Of course in the final formulation of Proposition 2.4 we lose some information, but the result will be applied to right K-invariant functions, and the left multiples of elements in $\tilde{\mathfrak{k}}$ annihilate such functions, so the result suffices for our purposes. Essentially we work on G/K.

We remind the reader of the direct image, see for instance [**JoL 01a**], Chapter II, §2. Suppose a manifold $\mathbf{X}$ is differentially isomorphic to a product $\mathbf{X}_1 \times \mathbf{X}_0$. Let π be the projection on $\mathbf{X}_1$, and let D be a differential operator on $\mathbf{X}$. Let x_0 be a given point of $\mathbf{X}_0$ and identify $\mathbf{X}_1$ with $\mathbf{X}_1 \times \{x_0\}$. We define the **direct image** $\pi_* D$ on $\mathbf{X}_1$ to be the differential operator such that for $f \in \mathrm{Fu}(\mathbf{X}_1)$,

$$(\pi_* D)f = (D(f \circ \pi))_{\mathbf{X}_1},$$

where the subscript $\mathbf{X}_1$ means restriction to $\mathbf{X}_1$. In practice for us, x_0 is the unit element in a group or coset space.

THEOREM 2.5. *Let $\omega = \omega_G$ be the Casimir operator on G.*
(i) *Let f be a function on G_P/K_{G_P}. Let $\pi : G/K \to G_P/K_{G_P}$ be the projection. For $u \in U_P, a \in A_P, g \in G_P$, we have*

$$(\omega(f \circ \pi))(uag) = (\omega_{G_P} f)(g).$$

In particular, $\pi_* \omega = \omega_{G_P}$.
(ii) *Let f be a function on A_P. Let*

$$H_{\tau_P} = \sum_{\alpha \in \mathcal{R}(\mathfrak{n}_P)} m(\alpha) H_\alpha \quad (\textit{and } m(\alpha) = 2 \textit{ on } \mathrm{SL}_n(\mathbf{C})).$$

Then

$$(\omega(f \circ \pi_{A_P}))(uag) = ((\omega_{A_P} - \tilde{H}_{\tau_P})f)(a).$$

In particular, $(\pi_{A_P})_* \omega = \omega_{A_P} - \tilde{H}_{\tau_P}$.

PROOF. This is a follow up of [**JoL 01a**], Chapter II, Lemma 3.2. Let f be a function on G_P/K_{G_P}, so $f \circ \pi$ is its right K-invariant lift to G. Then $f \circ \pi$ is left $U_P A_P$-invariant. In the decomposition of Theorem 2.1, we claim that both

$$\omega_{P\mathrm{unip}}(f \circ \pi)(uag) = \omega_{A_P}(f \circ \pi)(uag) = 0.$$

For simplicity, let us first check that if $Z \in \mathfrak{n}_P$ or $\mathfrak{a}_P$, then $\tilde{Z}(f \circ \pi)(uag) = 0$. Indeed, G_P normalizes $U_P A_P$, so for $g \in G_P$, we get

$$f(g \cdot \exp(tZ)) = f(\exp(tgZg^{-1})g) = f(g).$$

Taking d/dt yields 0. Let W be arbitrary in $\mathfrak{g}$. Then $\tilde{Z}\tilde{W}f(g)$ is computed from

$$f(g \cdot \exp(t_1 Z) \exp(t_2 W)).$$

We just saw that this expression is constant in t_1, so taking d/dt_1 yields 0, which proves our claim. Thus we have proved (i), and in particular

$$\pi_* \omega_G = \omega_{G_P}.$$

Similarly, let π be the projection on A_P. Let f be a function on A_P. Its pull back $f \circ \pi$ to G is U_P-invariant on the left and $G_P K$-invariant on the right. Now we use the fact that A_P normalizes U_P, and for $Z \in \mathfrak{n}_P$ (for instance $Z = E_\alpha$ or $\mathbf{i}E_\alpha$), and $a \in A_P$, we have $aZa^{-1} \in \mathfrak{n}_P$, so

$$f(a \cdot \exp(tZ)a^{-1}a) = f(\exp(taZa^{-1})a) = f(a).$$

Taking d/dt again yields 0. Arguing as before with the repeated derivative gives 0 also. Thus in the decomposition of Theorem 2.1 and Proposition 2.4, all the terms which don't come from differentiation $\tilde{H}$ with $H \in \mathfrak{a}_P$ are in the kernel of the projection. The remaining terms are precisely

$$\omega_{A_P} - \tilde{H}_{\tau_P},$$

which proves the theorem.

We shall apply the above decomposition to special types of functions which are decomposable in the following sense. Let:

f_{A_P} = a function on A_P lifted to G/K by the natural projection on A_P.
$h_{\mathbf{X}_{G_P}}$ = a function on G_P/K_{G_P} lifted to G/K by the natural projection on G_P/K_{G_P}.

Then the product $f_{A_P} h_{\mathbf{X}_{G_P}}$ is a function on G/K. For such a product we have a simplified expression for the action of ω, namely, a "Leibniz rule".

COROLLARY 2.6. *Notation as above, we have on $U_P A_P G_P$:*

$$\omega(f_{A_P} h_{\mathbf{X}_{G_P}}) = (\omega_{A_P} - \tilde{H}_{\tau_P}) f_{A_P} \cdot h_{\mathbf{X}_{G_P}} + f_{A_P} \cdot \omega_{G_P} h_{\mathbf{X}_{G_P}}.$$

PROOF. Immediate from Theorem 2.1, Proposition 2.4 and Theorem 2.5.

4.3. The differential equation for $E_{P,\mathbf{K}}$ and $E^{\#}_{P,\mathbf{K}}$

We work with the Casimir operator $\omega = \omega_G$ on G, actually on right K-invariant functions, so on G/K. By §2 we know how to apply it to a function on the various components of a parabolic, and in particular to functions on A_P and functions on G_P/K_{G_P}. We pull them back to G and apply ω, by definition of the direct image. Then we restrict back to A_P and G_P/K_{G_P} respectively.

Characters. Let $\chi = \chi_\zeta$ be a character on A_P, with $\zeta \in \mathfrak{a}^\vee_{P,\mathbf{C}}$. We view χ_ζ as a function on G/K via projection on A_P. We abbreviate the eigenvalue

$$(\mathrm{Iw}_{A_P})_*(\omega)\chi_\zeta = \omega\chi_\zeta = \mathrm{ev}(\omega, \chi_\zeta)\chi_\zeta.$$

Let $\langle \zeta, \eta \rangle = B^{\vee}_{\mathbf{C}}(\zeta, \eta)$ denote the $\mathbf{C}$-linear extension of $B_{\mathfrak{a}}$ to the complexified dual $\mathfrak{a}^{\vee}_{\mathbf{C}}$. By Theorem 2.5, the eigenvalue is given by

$$\text{(1)} \qquad \begin{aligned} \mathrm{ev}(\omega, \chi_\zeta) &= \langle \zeta, \zeta \rangle - \mathrm{ev}(\tilde{H}_{\tau_P}, \chi_\zeta) \\ &= \langle \zeta, \zeta \rangle - \zeta(H_{\tau_P}) = \langle \zeta, \zeta \rangle - \langle \zeta, \tau_P \rangle. \end{aligned}$$

Note that this eigenvalue is of the form $q^{\vee}(-\zeta)$ as in Chapter 2, Theorem 4.3 and Chapter 3, §4. The scalar product is the $\mathbf{C}$-bilinear symmetric product induced by the trace form on $\mathfrak{a}$. In any case, writing $\zeta = \xi + \mathbf{i}\lambda$ with $\xi, \lambda \in \mathfrak{a}^{\vee}_P$, we have

$$\langle \zeta, \zeta \rangle = -\langle \lambda, \lambda \rangle + 2\mathbf{i}\langle \xi, \lambda \rangle + \langle \xi, \xi \rangle.$$

We regard ξ as fixed. We then see that $\mathrm{ev}(\omega, \chi_{\xi+\mathbf{i}\lambda})$ has quadratic negative definite decay in λ, and thus for $t > 0$, the function

$$e^{\mathrm{ev}(\omega, \chi_\zeta)t}$$

has exponential quadratic decay in λ. With ξ, t fixed, this function is actually in the Gauss space of $\mathbf{i}\mathfrak{a}^{\vee}_P$. Thus it will present no convergence problem when we integrate it against bounded functions, or even functions with exponential linear growth on this imaginary space.

Let us write the abbreviation $\zeta^2 = \langle \zeta, \zeta \rangle$, and $\zeta \cdot \rho_P = \langle \zeta, \rho_P \rangle$, to get the formula in the classical notation

$$\mathrm{ev}(\omega, \chi_\zeta) = \zeta^2 - 2\zeta \cdot \rho_P = (\zeta - \rho_P)^2 - \rho_P^2.$$

This eigenvalue on χ_ζ will also turn out to be an eigenvalue on other functions, as we shall see immediately.

First we observe that the exponential function

$$e^{\mathrm{ev}(\omega, \chi_\zeta)t} = e^{(\zeta - \rho_P)^2 t} e^{-\rho_P^2 t}$$

has the above eigenvalue for the operator $\partial/\partial t$. Next we shall consider the Eisenstein series and the heat operator.

Heat equation. Let $\mathbf{K}_{\mathbf{X}_{G_P}}$ be the heat kernel on G_P/K_P. Let f_{A_P} be the function on A_P, lifted to G/K via the projection on A_P, defined by

$$f_{A_P,\zeta,\bar{\zeta},y}(x) = f_{A_P,\zeta,\bar{\zeta}}(x,y) = x^{\zeta}_{A_P} y^{\bar{\zeta}}_{A_P}.$$

Let

$$\text{(2)} \qquad \begin{aligned} F_P(\zeta, \bar{\zeta}, t, x, y) &= x^{\zeta}_{A_P} y^{\bar{\zeta}}_{A_P} \mathbf{K}_{\mathbf{X}_{G_P}}(t, x_{\mathbf{X}_{G_P}}, y_{\mathbf{X}_{G_P}}) \\ &= f_{A_P,\zeta,\bar{\zeta},y}(x) h_{\mathbf{X}_{G_P},y}(t,x) \end{aligned}$$

where $h_{\mathbf{X}_{G_P},y}(t,x) = \mathbf{K}_{\mathbf{X}_{G_P}}(t, x_{\mathbf{X}_{G_P}}, y_{\mathbf{X}_{G_P}})$ satisfies the heat equation on G_P/K_P. By Theorem 2.5(i), the function $h_{\mathbf{X}_{G_P},y}$ lifted to G satisfies the heat equation on G/K, and applying ω_{G_P} or ω_G to this function we get the same thing. However, with f_{A_P} and ω_{A_P}, ω_G there is an extra linear term by Theorem 2.5(ii). As a matter of notation we use throughout

$$\omega = \omega_{G,x} = \omega_x.$$

In the long run, to deal with positive rather than negative eigenvalues, we define

$$\boldsymbol{\Delta} = -\omega,$$

so $\boldsymbol{\Delta}$ is the (positive) Laplacian. The sign convention for positivity refers to operator positivity, in the sense that $\langle \boldsymbol{\Delta} f, f\rangle \geqq 0$, for the integral scalar product of functions on $\Gamma\backslash G/K$, hermitian in the second variable. We let

$$\mathbf{H} = \mathbf{H}_{x,t} = \boldsymbol{\Delta} + \partial_t = -\omega_x + \partial_t$$

be the **heat operator** on G/K.

PROPOSITION 3.1. *Let $\zeta \in \mathfrak{a}^{\vee}_{P,\mathbf{C}}$. The function on $\mathbf{R}_{>0} \times G/K$ given by*

$$(t,x) \mapsto F_P(\zeta, \bar{\zeta}, t, x, y) = F_{P,\zeta,\bar{\zeta},y}(t,x)$$

is an eigenfunction of the heat operator on G/K, with eigenvalue

$$\mathrm{ev}(\mathbf{H}_{x,t}, F_P(\zeta,\bar{\zeta},t,x,y)) = \mathrm{ev}(\boldsymbol{\Delta}, \chi_\zeta) = -\mathrm{ev}(\omega, \chi_\zeta).$$

PROOF. First take ∂_t. Then

$$\begin{aligned}\partial_t F_P(\zeta,\bar{\zeta},t,x,y) &= f_{A_P,\zeta,\bar{\zeta},y}(x)\omega_{G_P}\mathbf{K}\mathbf{x}_{G_P}(t, x\mathbf{x}_{G_P}, y\mathbf{x}_{G_P}) \\ &= (\omega_{G_P}F_P)(\zeta,\bar{\zeta},t,x,y).\end{aligned}$$

Next, using Corollary 2.6,

$$\begin{aligned}\omega_x F_P(\zeta,\bar{\zeta},t,x,y) =& (\omega_{A_P,x} - \tilde{H}_{\tau_P,x})(f_{A_P,\zeta,\bar{\zeta}}(x,y))\mathbf{K}\mathbf{x}_{G_P}(t, x\mathbf{x}_{G_P}, y\mathbf{x}_{G_P}) \\ &+ (\omega_{G_P}F_P)(\zeta,\bar{\zeta},t,x,y).\end{aligned}$$

Subtracting yields the proposition.

The eigenvalue of Proposition 3.1 will be called the **basic (P,ζ)-eigenvalue**, and will be denoted by $\mathrm{ev}_{P,\zeta}$, so

$$\boxed{\mathrm{ev}_{P,\zeta} = \mathrm{ev}(\mathbf{H}, F_{P,\zeta}) = \mathrm{ev}(\boldsymbol{\Delta}, \chi_\zeta) = -\mathrm{ev}(\omega, \chi_\zeta)}\,.$$

A general eigenvalue pattern. We shall now apply a general pattern as follows. Given any differential operator D on a manifold X, we can define the corresponding **heat operator** $\mathbf{H}_D$ to be

$$\mathbf{H}_D = D + \partial_t.$$

Let $F = F(t,x)$ be a function on the product $\mathbf{R}_{>0} \times X$, and suppose F is an eigenfunction of this heat operator with eigenvalue η. Then the function $F^{\#}$ given by

$$F^{\#}(t,x) = e^{-\eta t}F(t,x)$$

satisfies the heat equation $\mathbf{H}_D F^{\#} = 0$.

PROOF. Immediate.

In line with the above pattern, we define the function $F_P^{\#}$ by

$$F_P^{\#}(\zeta, \bar{\zeta}, t, x, y) = e^{-\mathrm{ev}_{P,\zeta}t} F_P(\zeta, \bar{\zeta}, t, x, y). \tag{3}$$

We apply the general pattern to Proposition 3.1 to get:

PROPOSITION 3.2. *The function* $(t, x) \mapsto F_P^{\#}(\zeta, \bar{\zeta}, t, x, y)$ *satisfies the heat equation.*

The function F_P in (2) of course is used to define the terms of the **heat Eisenstein series**, namely

$$\begin{aligned} &E_P(\mathrm{Tr}_{\Gamma_{G_P}}(\mathbf{K}_{\mathbf{X}_{G_P}}), \chi_\zeta, \bar{\chi}_\zeta)(t, x, y) \\ &\quad = \sum_{\gamma_1, \gamma_2 \in \Gamma_P \backslash \Gamma} \mathrm{Tr}_{\Gamma_{G_P}}(F_P)(\zeta, \bar{\zeta}, t, \gamma_1 x, \gamma_2 y). \end{aligned}$$

The variables (ζ, t, x, y) are now going to play a role simultaneously, so we shall use the notation

$$\boxed{E_P(\mathrm{Tr}_{\Gamma_{G_P}}(\mathbf{K}_{\mathbf{X}_{G_P}}), \chi_\zeta, \bar{\chi}_\zeta)(t, x, y) = E_{P,\Gamma,\mathbf{K}}(\zeta, \bar{\zeta}, t, x, y)}.$$

PROPOSITION 3.3. *The Eisenstein series* $E_{P,\Gamma,\mathbf{K},\zeta}$ *is an eigenfunction of the heat operator, with eigenvalue* $\mathrm{ev}_{P,\zeta}$ *for* $\mathrm{Re}(\zeta) > 2\rho_P$.

PROOF. Each term of the series for $E_{P,\Gamma,\mathbf{K},\zeta,\bar{\zeta}}$ is an eigenfunction with the stated eigenvalue by Proposition 3.1. We can differentiate term by term, so the proposition is immediate.

Just as for $F_P^{\#}$ we define the **heated Eisenstein series**

$$E_{P,\Gamma,\mathbf{K}}^{\#}(\zeta, \bar{\zeta}, t, x, y) = e^{-\mathrm{ev}_{P,\zeta}t} E_{P,\Gamma,\mathbf{K}}(\zeta, \bar{\zeta}, t, x, y). \tag{4}$$

PROPOSITION 3.4. *Let* $\mathrm{Re}(\zeta) = \xi > 2\rho_P$. *Then*

$$E_{P,\Gamma,\mathbf{K}}^{\#}(\zeta, \bar{\zeta}, t, x, y)$$

satisfies the heat equation.

PROOF. Special case of the general eigenvalue pattern before Proposition 3.2.

We then consider the kernel function $J_{P,\Gamma,\xi,t}$ of Chapter 3, §4, given by the integral

$$J_{P,\Gamma,\xi,t}(x, y) = \int_{\mathrm{Re}(\zeta)=\xi} E_{P,\Gamma,\mathbf{K}}^{\#}(\zeta, \bar{\zeta}, t, x, y) d\mathrm{Im}(\zeta). \tag{5}$$

The corresponding operator will be called the (P,ξ)-**heated Eisenstein operator**, and its kernel function $J_{P,\Gamma,\xi,t}$ will be called the (P,ξ)-**heated Eisenstein kernel**. The function J is Γ-invariant since the Eisenstein series of the integrand is Γ-invariant. So in the (x,y) variables, it is defined on $\Gamma\backslash G \times \Gamma\backslash G$.

PROPOSITION 3.5. *Let* $\mathrm{Re}(\zeta) = \xi > 2\rho_P$. *Then* $J_{P,\Gamma,\xi,t}$ *satisfies the heat equation.*

PROOF. This comes from Proposition 3.4 and differentiation under the integral sign.

The Gauss space Fourier inversion does not appear explicitly in the above proposition. It will appear in §5, so we make a few additional remarks. As in Chapter 3, §4, we let φ_t be the function in the Gauss space such that

$$(\mathbf{M}\varphi_t)(-\zeta) = e^{\mathrm{ev}(\omega,\chi_\zeta)t} = e^{-\mathrm{ev}_{P,\zeta}t}.$$

As remarked at the beginning of this section, in the notation of Chapter 3, §4,

$$\mathrm{ev}(\omega,\chi_\zeta) = q^\vee(-\zeta) = \langle\zeta,\zeta\rangle - \langle\zeta,\tau_P\rangle.$$

Thus the J-function can also be expressed in the form

$$J_{P,\Gamma,\xi,t}(x,y) = ((\mathbf{M}\varphi_t)^- *_\xi E_{P,\Gamma,\mathbf{K},t})(x,y). \tag{6}$$

This expression will be relevant in §5 below.

4.4. Convolution of $\mathrm{Tr}_\Gamma(\mathbf{K_X})$ and the Eisenstein series

Let $\mathbf{X} = G/K$ with $G = \mathrm{SL}_n(\mathbf{C})$ for concreteness. Let $\mathbf{K_X}$ be the heat kernel on $\mathbf{X}$. As described at the end of Chapter 2, §2, it can be expressed by a formula which shows that it is symmetric. As before, we let

$$\Gamma = \Gamma_n = \mathrm{SL}_n(\mathbf{Z}[\mathbf{i}]).$$

We consider the Γ-trace of the heat kernel

$$\mathrm{Tr}_\Gamma(\mathbf{K_X})(t,x,y) = \sum_{\gamma\in\Gamma} \mathbf{K_X}(t,\gamma x,y).$$

From the formula via the Gaussian $\mathbf{g}_t$, we have already seen in Chapter 2,§2 that this trace is symmetric in (x,y). It is Γ-invariant. We may view $\mathrm{Tr}_\Gamma(\mathbf{K_X})$ as a function on $\mathbf{R}_{>0} \times (\Gamma\backslash\mathbf{X}) \times (\Gamma\backslash\mathbf{X})$. However, $\Gamma\backslash\mathbf{X}$ may have singularities, so we view this trace as a function on $\Gamma\backslash G$, which is also right K-invariant, or as a function on G which is left Γ-invariant and right K-invariant, or as we also say, (Γ,K)-invariant.

We use the Dodziuk criterion reproduced as Theorem 1.1 of the Appendix. The remarks following this theorem show how the uniqueness theorem applies in the present context. Thus we obtain:

THEOREM 4.1. *Let $h(t,x)$ be a solution of the heat equation on G/K, Γ-invariant, bounded continuous and initially complete. Then h is uniquely determined by its initial condition.*

PROOF. Dodziuk's theorem, see the Appendix, Theorem 1.1.

On $\Gamma\backslash G$ we can convolve the heat kernel with a function h as above, namely

$$(\mathrm{Tr}_\Gamma(\mathbf{K}_{\mathbf{X},t_1}) * h_{t_2})(x) = \int_{\Gamma\backslash G} \mathrm{Tr}_\Gamma(\mathbf{K}_\mathbf{X})(t_1,x,y)h(t_2,y)dy.$$

In particular, **by definition**, for $x, z \in G, \zeta \in \mathfrak{a}^\vee_{P,\mathbf{C}}$ and $\mathrm{Re}(\zeta)$ sufficiently large,

$$\begin{aligned}(1)\qquad &\mathrm{Tr}_\Gamma(\mathbf{K}_{\mathbf{X},t_1}) * E^\#_{P,\Gamma,\mathbf{K},t_2}(\zeta,\bar{\zeta},x,z)\\ &= \int_{\Gamma\backslash G} \mathrm{Tr}_\Gamma(\mathbf{K}_{\mathbf{X},t_1})(x,y)E^\#_{P,\Gamma,\mathbf{K},t_2}(\zeta,\bar{\zeta},y,z)dy.\end{aligned}$$

THEOREM 4.2. *We have the equality for $t_1, t_2 > 0, \mathrm{Re}(\zeta) > 2\rho_P$,*

$$\mathrm{Tr}_\Gamma(\mathbf{K}_{\mathbf{X},t_1}) * E^\#_{P,\Gamma,\mathbf{K},t_2}(\zeta,\bar{\zeta},x,z) = E^\#_{P,\Gamma,\mathbf{K}}(\zeta,\bar{\zeta},t_1+t_2,x,z).$$

PROOF. We take $-\omega_x + \partial_{t_1}$ under the integral sign in (1), and get 0 by the fact that $\mathbf{K}_\mathbf{X}$ satisfies the heat equation. Letting $t_1 \to 0$ and using the Dirac property of $\mathrm{Tr}_\Gamma(\mathbf{K}_\mathbf{X})$ for bounded (Γ, K)-invariant continuous functions shows that the convolution on the left of (1) has the initial condition

$$E^\#_{P,\Gamma,\mathbf{K}}(\zeta,\bar{\zeta},t_2,x,z).$$

On the other hand, the function $E^\#_{P,\Gamma,\mathbf{K}}(\zeta,t_1+t_2,x,z)$ satisfies the heat equation by Proposition 3.4, and has the same initial condition. By Theorem 4.1, this concludes the proof of Theorem 4.2.

4.5. The P-anticuspidal semigroup property

This section starts the study of the extent to which $J_{P,\Gamma,\xi,t}$ defines a semigroup of operators. For ξ sufficiently large, so that the Eisenstein series converge, only a perturbation of the semigroup property is satisfied. It will take considerably more work to provide an analytic continuation in the variable ξ, and to show that for the special value $\xi = \rho_P$, the **semigroup property** is satisfied; that is, with the normalization $\mathrm{vol}(\Gamma_{U_P}\backslash U_P) = 1$, we expect

$$J_{P,\Gamma,\rho_P,t_1} *_{\Gamma\backslash G} J_{P,\Gamma,\rho_P,t_2} = J_{P,\Gamma,\rho_P,t_1+t_2}.$$

Actually, we are trying to prove something even stronger, namely:

CONJECTURE 5.1. *The Eisenstein series* $E^{\#}_{P,\Gamma,K}(\zeta_1, \zeta_2, t, x, y)$ *(which has two zeta variables) can be continued to a meromorphic function in both variables. Putting* $\zeta_2 = \bar{\zeta}_1$*, the kernel function* $J_{P,\Gamma,\xi,t}$ *can be continued to a real analytic function on the space* $\mathrm{Re}(\zeta) = \rho_P$*. Furthermore, for*

$$\xi = \rho_P = \mathrm{Re}(\zeta_1),$$

we have

$$\int_{\Gamma\backslash G} E^{\#}_{P,\Gamma,\mathbf{K}}(\zeta_1, \bar{\zeta}_1, t_1, x, y) J_{P,\Gamma,\rho_P}(t_2, y, z) dy$$
$$= \mathrm{vol}(\Gamma_{U_P}\backslash U_P) E^{\#}_{P,\Gamma,\mathbf{K}}(\zeta_1, \bar{\zeta}_1, t_1 + t_2, x, z).$$

Or writing the inner integral more explicitly,

$$\int_{\Gamma\backslash G} E^{\#}_{P,\Gamma,\mathbf{K}}(\zeta_1, \bar{\zeta}_1, t_1, x, y) \int_{\mathrm{Re}(\zeta)=\rho_P} E^{\#}_{P,\Gamma,\mathbf{K}}(\zeta, \bar{\zeta}_1, t_2, y, z) d\mathrm{Im}(\zeta) dy$$
$$= \mathrm{vol}(\Gamma_{U_P}\backslash U_P) E^{\#}_{P,\Gamma,\mathbf{K}}(\zeta_1, \bar{\zeta}_1, t_1 + t_2, x, z).$$

Or in the convolution notation eliminating the integral signs, for $\mathrm{Re}(\zeta_1) = \rho_P$*,*

$$E^{\#}_{P,\Gamma,\mathbf{K},t_1} * J_{P,\Gamma,\rho_P,t_2} = \mathrm{vol}(\Gamma_{U_P}\backslash U_P) E^{\#}_{P,\Gamma,\mathbf{K},t_1+t_2}.$$

In this section, we only give a formula for $\xi > 2\rho_P$, in which case we meet Eisenstein series in two complex characters and some translations.

As in §3, and Chapter 3, §4, we shall write any one of the three expressions

$$e^{q^{\vee}(-\zeta_1)t_2} = e^{\mathrm{ev}(\omega_G, \chi_{\zeta_1})t_2} = \mathbf{M}\varphi_{t_2}(-\zeta_1).$$

For $\xi = \rho_P = \mathrm{Re}(\zeta_1)$, the formula in Conjecture 5.1 amounts to

$$\text{(1)} \quad \int_{\Gamma\backslash G} E_{P,\Gamma,\mathbf{K}}(\zeta_1, \bar{\zeta}_1, t_1, x, y) J_{P,\Gamma,\rho_P}(t_2, y, z) dy$$
$$= \mathrm{vol}(\Gamma_{U_P}\backslash U_P) e^{q^{\vee}(-\zeta_1)t_2} E_{P,\Gamma,\mathbf{K}}(\zeta_1, \bar{\zeta}_1, t_1 + t_2, x, z).$$

Without analytic continuation, one obtains only a perturbation. To state it, we recall the **two characters Eisenstein series**, with two variables $\zeta_1, \zeta_2 \in \mathfrak{a}^{\vee}_{P,\mathbf{C}}$, namely

$$E_{P,\Gamma,\mathbf{K}}(\zeta_1, \zeta_2, t, x, z)$$
$$= \sum_{\gamma,\gamma' \in \Gamma_P\backslash\Gamma} \mathrm{Tr}_{\Gamma_{G_P}}(\mathbf{K}_{\mathbf{X}_{G_P}})(t, (\gamma x)_{\mathbf{X}_{G_P}}, (\gamma' z)_{\mathbf{X}_{G_P}}) (\gamma x)^{\zeta_1}_{A_P} (\gamma' z)^{\zeta_2}_{A_P}.$$

THEOREM 5.2. *For ξ and $\mathrm{Re}(\zeta_1) > 2\rho_P$, we have the limit*

$$\lim_{t_1 \to 0} \int_{\Gamma\backslash G} E_{P,\Gamma,\mathbf{K}}(\zeta_1, \bar{\zeta}_1, t_1, x, y) J_{P,\Gamma,\xi}(t_2, y, z) dy$$

$$= \mathrm{vol}(\Gamma_{U_P}\backslash U_P)\mathbf{M}\varphi_{t_2}(\bar{\zeta}_1 - 2\rho_P) E_{P,\Gamma,\mathbf{K}}(\zeta_1, \bar{\zeta}_1 + 2\xi - 2\rho_P, t_2, x, z)$$

$$+ \text{ a term given in (7) below, with } \gamma_3 \neq \mathrm{id}.$$

Note that if we set formally $2\xi - 2\rho_P = 0$, that is $\xi = \rho_P$ and let $\mathrm{Re}(\zeta_1) = \xi$, then this limit is equal to the limit of the right side of (1), namely

$$\mathrm{vol}(\Gamma_{U_P}\backslash U_P)\mathbf{M}\varphi_{t_2}(-\zeta_1) E_{P,\Gamma,\mathbf{K}}(\zeta_1, \bar{\zeta}_1, t_2, x, z),$$

plus the extra term as in (7). Indeed, putting $\zeta_1 = \xi + \mathbf{i}\lambda_1$, we have

$$\zeta_1 - 2\rho_P = -\zeta_1,$$

so the desired expression drops out. However, the Eisenstein series with such values for $\mathrm{Re}(\zeta_1)$ and $\mathrm{Re}(\zeta)$ do not converge, so there is a need to develop a whole theory of analytic continuation to get the simpler one-parameter semigroup formula. The analytic continuation should hold for the two-variable Eisenstein series, and then one puts $\zeta_2 = \bar{\zeta}_1$ as needed.

We now embark on the proof of Theorem 5.2. We first apply Chapter 3, §2, formula (6). The function ψ is taken to be (for $\xi > 2\rho_P$)

$$\psi(y) = J_{P,\Gamma,\xi}(t_2, y, z)$$

$$(2) \qquad = \sum_{\gamma_3,\gamma_4} \mathrm{Tr}_{\Gamma_{G_P}}(\mathbf{K}_{\mathbf{X}_{G_P}})(t_2, (\gamma_3 y)_{\mathbf{X}_{G_P}}, (\gamma_4 z)_{\mathbf{X}_{G_P}}) \cdot$$

$$\cdot \int_{\mathrm{Re}(\zeta)=\xi} (\gamma_3 y)^{\zeta}_{A_P} (\gamma_4 z)^{\bar{\zeta}}_{A_P} \mathbf{M}\varphi_{t_2}(-\zeta) d\mathrm{Im}(\zeta).$$

Let $\mathrm{LS}(\xi, t_1)$ be the left side of (1) with $\xi > 2\rho_P$ instead of $\xi = \rho_P$. Thus

$$\mathrm{LS}(\xi, t_1) = E_{P,\Gamma,\mathbf{K},t_1}(\zeta_1, \bar{\zeta}_1) * J_{P,\Gamma,\xi,t_2}.$$

Chapter 3, §2 (6) (see also Chapter 2, Proposition 5.2) yields

$$(3) \qquad \mathrm{LS}(\xi, t_1) = \sum_{\gamma_1} (\gamma_1 x)^{\zeta_1}_{A_P} \int_{\Gamma_{U_P}\backslash U_P} \int_{A_P} \int_{\Gamma_{G_P}\backslash G_P} F(u, a, g) du da dg$$

where

$$F(u, a, g) = \mathrm{Tr}_{\Gamma_{G_P}}(\mathbf{K}_{\mathbf{X}_{G_P}})(t_1, (\gamma_1 x)_{\mathbf{X}_{G_P}}, g) a^{\bar{\zeta}_1 - 2\rho_P} \psi(uag).$$

Thus we substitute uag for y in (2), occurring in two places $(\gamma_3 uag)_{\mathbf{X}_{G_P}}$ and $(\gamma_3 uag)_{A_P}$. We let $t_1 \to 0$. We note the inner integral which is a convolution with respect to the variable $g \in G_P$ of the heat kernel on $\Gamma_{G_P}\backslash G_P/K_{G_P}$, and the function $\psi(uag)$ as function of g. By the Dirac property of the heat kernel, we obtain

$$\lim_{t_1\to 0} \mathrm{LS}(\xi,t_1) = \sum_{\gamma_1}(\gamma_1 x)_{A_P}^{\zeta_1} \int_{\Gamma_{U_P}\backslash U_P}\int_{A_P} \psi(ua(\gamma_1 x)\mathbf{x}_{G_P})a^{\bar{\zeta}_1-2\rho_P}duda$$

$$(4) \qquad = \sum_{\gamma_1,\gamma_3,\gamma_4}(\gamma_1 x)_{A_P}^{\zeta_1} \int_{\Gamma_{U_P}\backslash U_P}\int_{A_P} I_\xi(\gamma_1,\gamma_3,\gamma_4,u,a)duda$$

where

$$(5) \qquad I_{\xi,t_2}(\gamma_1,\gamma_3,\gamma_4,u,a) =$$
$$\mathrm{Tr}_{\Gamma_{G_P}}(\mathbf{K}\mathbf{x}_{G_P})(t_2,(\gamma_3 ua(\gamma_1 x)x_{G_P},(\gamma_4 z)x_{G_P})\cdot$$
$$\cdot\, a^{\bar{\zeta}_1-2\rho_P}\int_{\mathrm{Re}(\zeta)=\xi}(\gamma_3 ua(\gamma_1 x)x_{G_P})_{A_P}^{\zeta}(\gamma_4 z)_{A_P}^{\bar{\zeta}}\mathbf{M}\varphi_{t_2}(-\zeta)d\mathrm{Im}(\zeta).$$

We have to evaluate (5). We distinguish the two cases when $\gamma_3 = \mathrm{id}$ and $\gamma_3 \neq \mathrm{id}$.

Suppose first $\gamma_3 = \mathrm{id}$. Then

$$(6) \qquad I_{\xi,t_2}(\gamma_1,\mathrm{id},\gamma_4,u,a) =$$
$$\mathrm{Tr}_{\Gamma_{G_P}}(\mathbf{K}\mathbf{x}_{G_P})(t_2,(\gamma_1 x)x_{G_P},(\gamma_4 z)x_{G_P})\cdot$$
$$\cdot\, a^{\bar{\zeta}_1-2\rho_P}\int_{\mathrm{Re}(\zeta)=\xi} a^{\zeta}(\gamma_4 z)_{A_P}^{\bar{\zeta}}\mathbf{M}\varphi_{t_2}(-\zeta)d\mathrm{Im}(\zeta).$$

We note that this expression is independent of u. Integrating over $\Gamma_{U_P}\backslash U_P$ simply multiplies the rest by the factor $\mathrm{vol}(\Gamma_{U_P}\backslash U_P)$. This leaves the integral over A_P of the expression in (6). The first part involving the heat kernel on $\mathbf{X}_{G_P}$ is independent of the variable a. Thus we are left with two integrals, over $d\mathrm{Im}(\zeta)$ and over da respectively, which are routinely evaluated as follows as a manifestation of Fourier-Mellin inversion.

LEMMA 5.3. *Let $b \in A_P$. In the application, $b = (\gamma_4 z)_{A_P}$. Let $\varphi \in \mathrm{Gauss}(\mathfrak{a}_P)$. Then:*

$$(\mathrm{i}) \qquad \int_{\mathrm{Re}(\zeta)=\xi} a^{\zeta}b^{\bar{\zeta}}\mathbf{M}\varphi(-\zeta)d\mathrm{Im}(\zeta) = b^{2\xi}\varphi(a/b).$$

$$(\mathrm{ii}) \qquad \int_{A_P} a^{\bar{\zeta}_1-2\rho_P}b^{2\xi}\varphi(a/b)da = b^{\bar{\zeta}_1-2\rho_P}b^{2\xi}\mathbf{M}\varphi(\bar{\zeta}_1-2\rho_P).$$

PROOF. Each part is immediate from Mellin inversion. For (i) we write $\zeta = \xi + \mathbf{i}\lambda$, and take $a^{\xi}b^{\xi}$ out of the integral. Cf. Chapter 3, §2, (1) with $\eta = \xi$. For (ii), we make the multiplicative translation $a \mapsto ab$ and apply the definition of the Mellin transform.

If we plug in the value of Lemma 5.3(ii) into the right side of (4), we find that we have proved the following formula.

$$(7)\qquad \lim_{t_1\to 0}\int_{\Gamma\backslash G} E_{P,\Gamma,\mathbf{K}}(\zeta_1,\bar{\zeta}_1,t_1,x,y)J_{P,\Gamma,\xi}(t_2,y,z)dy =$$

$$\mathrm{vol}(\Gamma_{U_P}\backslash U_P)\mathbf{M}\varphi_{t_2}(\bar{\zeta}_1-2\rho_P)E_{P,\Gamma,\mathbf{K}}(\zeta_1,\bar{\zeta}_1+2\xi-2\rho_P,t_2,x,z)$$

$$+\sum_{\gamma_1,\gamma_3\neq \mathrm{id},\gamma_4}(\gamma_1 x)_{A_P}^{\zeta_1}\int_{\Gamma_{U_P}\backslash U_P}\int_{A_P} I_{\xi,t_2}(\gamma_1,\gamma_3,\gamma_4,u,a)duda$$

The first term on the right in this formula is exactly the expression given in the statement of Theorem 5.2, and proves the theorem.

It is now a problem to show that the last sum is 0 for $\xi=\rho_P$. Just to make sense, it requires some continuation of the Heat Eisenstein series to the vertical space $\mathrm{Re}(\zeta)=\rho_P$. Then a proof could be given for Conjecture 5.1 in the same way as for Theorem 4.2, because of the heat equation and the determination of the initial conditions.

4.6. The P-anticuspidal operator J_{P,Γ,ρ_P} and the conjectured spectral expansion

In Chapter 3, Theorem 3.5, we obtained a condition for any $\psi\in \mathrm{BC}(\Gamma\backslash G/K)$ to be P-cuspidal, in terms of convolution with the heat kernel. We shall develop a condition conjecturally at the center of the critical strip. We define $L^2_{\mathrm{dis}}(\Gamma\backslash G/K)$ to be the closure of the subspace of L^2 generated by eigenfunctions of Casimir, and call this subspace the **discrete part of** L^2. We define a function to be **cuspidal** if it is P-cuspidal for all P, and call the space of such functions the **cuspidal space**. It is contained in the discrete part of L^2. Its orthogonal complement will be called the **residual space** $L^2_{\mathrm{res}}(\Gamma\backslash G/K)$.

For $n=2$, the discrete part of L^2 is the cuspidal subspace plus the constants. For $n>2$, it involves more, namely a bigger residual part, for which the only available basic references is the "jungle" of Langlands [**Lglds 76**], §7. Gelfand-Piatecki-Shapiro proved that convolution on the cuspidal space with an L^1 function is a compact operator. Borel-Garland [**BoG82**] extended this to the full discrete part.

CONJECTURE 6.1. *Suppose that for all ζ with* $\mathrm{Re}(\zeta)=\rho_P$, $t>0$, $x\in G$, *we have*

$$(E_{P,\Gamma,\mathbf{K}}*\psi)(t,\zeta,\bar{\zeta},x)=0.$$

Then ψ is in the P-cuspidal + discrete subspace.

The convolution product is on $\Gamma\backslash G$. The integral implicit in this product is taken over the second $\Gamma\backslash G$-variable.

One question is of course for which functions ψ does the conclusion hold, that is, one must eventually make precise a vector space of functions. The condition

C^∞ with compact support is not good enough, because the space of functions must include the heat kernel as a top priority. The condition of C^∞ and bounded $\mathrm{BC}^\infty(\Gamma\backslash G/K)$ for the function and all its derivatives may be sufficient for all our purposes. It is satisfied by the heat kernel, going back to Chapter 1, Lemma 2.4 and Chapter 2, §2 which gave the explicit formula for the heat kernel. Even more appropriately, we propose a new space of test functions, the gaussians, amounting to linear combinations of the heat gaussian with different values of t; see [**JoL 04a**].

The function $J_{P,\Gamma,\xi,t}$ has previously reflected at least two formalisms, namely the eigenvalue formalism of the general pattern starting with Proposition 3.2, and a Mellin inversion formalism, where it was relevant to use the Mellin transform $\mathbf{M}\varphi_t$. The present section deals exclusively with the eigenvalue formalism. Furthermore, for the purposes of this section, we go even beyond in that we assume analytic continuation, and use $\xi = \rho_P$ at the center of the (higher dimensional) critical strip. Thus we deal with

$$(1)\qquad \begin{aligned} J_{P,\Gamma,\rho_P,t}(x,z) &= \int\limits_{\mathrm{Re}(\zeta)=\rho_P} E^{\#}_{P,\Gamma,\mathbf{K}}(t,\zeta,\bar{\zeta}x,z)d\mathrm{Im}(\zeta) \\ &= \int\limits_{\mathrm{Re}(\zeta)=\rho_P} e^{-\mathrm{ev}_{P,\zeta}t}E_{P,\Gamma,\mathbf{K}}(t,\zeta,\bar{\zeta},x,z)d\mathrm{Im}(\zeta). \end{aligned}$$

As a formal consequence of Conjecture 5.1 and 6.1, we obtain:

COROLLARY 6.2. *The map $t \to J_{P,\Gamma,K,\rho_P,t}(x,z)$ satisfies the semigroup property under convolution on $\Gamma\backslash G$. For ψ in the appropriate space (including the heat kernel) the function*

$$\psi - \lim_{t\to 0} J_{P,\Gamma,\rho_P,t} * \psi$$

is in the P-cuspidal subspace + residual subspace. The measures are assumed normalized so that $\mathrm{vol}(\Gamma_{U_P}\backslash U_P) = 1$.

FORMAL PROOF. Let $\mathrm{Re}(\zeta_1) = \rho_P$. We write down only the indices relevant to the proof, so P, ζ_1 and t. We apply the convolution operator with $E^{\#}_{P,\zeta_1,t_1}$ to the given function, convolution being the integral over $\Gamma\backslash G$ in the last variable. We obtain:

$$\begin{aligned} &E^{\#}_{P,\zeta_1,t_1} * (\psi - \lim_{t_2\to 0} J_{P,\rho_P,t_2} * \psi) \\ &\qquad = E^{\#}_{P,\zeta_1,t_1} * \psi - \lim_{t_2\to 0} E^{\#}_{P,\zeta_1,t_1} * J_{P,\rho_P,t_2} * \psi \\ &\qquad = E^{\#}_{P,\zeta_1,t_1} * \psi - \lim_{t_2\to 0} E^{\#}_{P,\zeta_1,t_1+t_2} * \psi \quad \text{[by Conjecture 5.1]} \\ &\qquad = E^{\#}_{P,\zeta_1,t_1} * \psi - E^{\#}_{P,\zeta_1,t_1} * \psi \\ &\qquad = 0. \end{aligned}$$

Then Conjecture 6.1 concludes the formal proof.

The analytic continuation of the other part of Chapter 3, Theorem 3.5 would show in addition that for any test function ψ_1 which is P-cuspidal, and ψ as in Corollary 6.2, the function $J_P * \psi$ is orthogonal to ψ_1 (on $\Gamma\backslash G$). This simply comes formally from Fubini and the associativity of convolutions.

For purely combinatorial reasons, taking the convolution product of enough factors of type $(I - J_P)$ with P ranging over the reduced parabolics, Corollary 6.2 is accompanied by the following simultaneous cuspidalization.

CONJECTURE 6.3. *There exist real numbers c_P such that for all ψ, the function*

$$\psi - \sum_P c_P \lim_{t\to 0} J_{P,\Gamma,\rho_P,t}(\psi)$$

is in the cuspidal + residual = discrete subspace.

The significance of Conjecture 6.3 is that convolution with a function which has sufficiently fast decay and is cuspidal is a compact operator, which therefore has an ordinary Fourier type series.

Having this extension, one proceeds as follows.

Let $\{\psi_k\}$ be an orthonormal basis for the discrete part of L^2, consisting of square integrable eigenfunctions for $\mathbf{\Delta}$, with eigenvalues λ_k, so that

$$\mathbf{\Delta}\psi_k = \lambda_k \psi_k.$$

For each P, we have several objects associated with P, and we index them systematically with P. In addition to $E_{P,\Gamma,\mathbf{K},\zeta}, E_P^{\#}, J_P$, we index $\chi_{P,\zeta} = \chi_\zeta$ for $\zeta \in \mathfrak{a}_P^\vee$. Let $\mathbf{H}$ be the heat operator. From Theorems 3.1 and 3.3, we know that the Eisenstein series $E_{P,\Gamma,\mathbf{K},\zeta}$ is an eigenfunction of $\mathbf{H}$, with the basic eigenvalue

$$\mathrm{ev}_{P,\zeta} = \mathrm{ev}(\mathbf{H}, E_{P,\Gamma,\mathbf{K},\zeta}) = \mathrm{ev}(\mathbf{\Delta}, \chi_{P,\zeta}).$$

Then we expect the eigenfunction decomposition of the heat kernel as follows. There exists a real number c'_P for each reduced parabolic P, and a normalization of the Haar measures (giving appropriate values to the volumes of $\Gamma\backslash G, \Gamma_{U_P}\backslash U_P$, etc.), such that for the heat kernel itself we have:

EFEX 1. $\quad \mathrm{Tr}_\Gamma(\mathbf{K_X})(t,x,z) =$

$$\sum_k \psi_k(x)\overline{\psi_k(z)}e^{-\lambda_k t} + \sum_P c'_P \int_{\mathrm{Re}(\zeta)=\rho_P} e^{-\mathrm{ev}_{P,\zeta} t} E_{P,\Gamma,\mathbf{K}}(t,\zeta,\bar{\zeta},x,z)d\mathrm{Im}(\zeta).$$

The above expansion is a mixture of a series analogous to a Fourier series, and a continuous part analogous to what is needed for the Fourier transform. Here we are faced with "P-Eisenstein transforms". We chose the expression for $J_{P,\Gamma,\rho_P,t}$ which has the same formal structure as the terms of the Fourier-like series coming first. It may be suggestive to call the integral a theta integral in lieu of a theta series ...

Having such an expansion for the heat kernel, one can then use the Dirac property of the heat kernel to get an expansion for more general functions, for instance for a function φ, sufficiently rapidly decreasing on $\Gamma\backslash G/K$:

EFEX 2. $\varphi(x) =$

$$\sum_k \langle \varphi, \psi_k \rangle \psi_k(x) + \sum_P c'_P \lim_{t \to 0} \int_{\mathrm{Re}(\zeta)=\rho_P} (E^{\#}_{P,\Gamma,\mathbf{K}} * \varphi)(t, \zeta, \bar{\zeta}, x) d\mathrm{Im}(\zeta).$$

The convolution in the integrand is the scalar product in the last variable, that is,

$$(E^{\#} * \varphi)(t, \zeta, \bar{\zeta}, x) = \int_{\Gamma \backslash G} E^{\#}(t, \zeta, \bar{\zeta}, x, z) \varphi(z) dz_{\Gamma \backslash G}.$$

Note that since one is taking the limit as t approaches zero, one can work with E instead of $E^{\#}$ and the formula remains valid. Following the context of Chapter 3, §4, (2b), reversing the order of integration, this expression can also be written in the form

$$\varphi(x) = \sum_k \langle \varphi, \psi_k \rangle \psi_k(x) + \sum_P c'_P \lim_{t \to 0} J_{P,\Gamma,\rho_P,t}(\varphi)(x).$$

This way of writing reflects the P-anti-discrete property of the integral operator represented by J_P. The sum on the right with appropriate numbers is a continuous projection of the function φ, which after subtraction admits an ordinary eigenfunction series of the form associated to the compact case.

At the moment, the main obstacle to do all this lies in the analytic continuation of the heat Eisenstein series in the variable ζ.

Special Case. In certain special cases **EFEX 1** can be compared to existing results. Let P be the minimal reduced parabolic, corresponding to the partition of n with $n_i = 1$ for all i. The group G_{n_i} is simply a point, and the defining characterization of the heat kernel is such that the only reasonable definition for $\mathbf{K}$ on G_{n_i} is the constant function 1. Thus $\mathbf{K}_{\mathbf{X}_{G_P}} = 1$. By Remark 2 of Chapter 2, §3, we get

$$E_{P,\Gamma,\mathbf{K}}(\zeta, \bar{\zeta}, t, x, y) = E_P(\chi_\zeta)(x) E_P(\chi_{\bar{\zeta}})(y)$$

where $E_P(\chi_\zeta)$ is the character Eisenstein series defined in Chapter 2, §1. The meromorphic continuation of $E_P(\chi_\zeta)(x)$ is known (see Maass [**Maa 71**], Borel [**Bor 97**], [**JoL 01b**] and [**JoL 04**]). Consequently, the term in the conjectured expansion **EFEX 1** corresponding to the minimal parabolic exists at $\mathrm{Re}(\zeta) = \rho_P$ and, indeed, the integral

$$\int_{\mathrm{Re}(\zeta)=\rho_P} e^{-\mathrm{ev}_{P,\zeta} t} E_P(\chi_\zeta)(x) E_P(\chi_{\bar{\zeta}})(z) d\mathrm{Im}(\zeta)$$

exists for all x and z.

On SL_2, the only standard parabolic is the minimal parabolic. For $\mathrm{SL}_2(\mathbf{C})$, the expansion in **EFEX 1** becomes

$$\mathrm{Tr}_\Gamma(\mathbf{K})(t, x, z) =$$

$$\sum_k \psi_k(x) \overline{\psi_k(z)} e^{-\lambda_k t} + \int_{\mathrm{Re}(\zeta)=1} e^{\zeta(\zeta-2)t} E_P(\chi_\zeta)(x) E_P(\chi_{\bar{\zeta}})(z) d\mathrm{Im}(\zeta),$$

which coincides with the known spectral expansion of the heat kernel on the quotient space $\mathrm{SL}_2(\mathbf{Z}[\mathbf{i}]) \backslash \mathrm{SL}_2(\mathbf{C})$. For $\mathrm{SL}_2(\mathbf{R})$, which is entirely similar, the formula in the context of functional analysis is apparently due to Roelcke as proved in [**Roe 66**], [**Roe 67**], and referred to in Kubota [**Kub 74**] p. 62. For another idea leading to the spectral expansion of an entirely different kind of test function and its application to the trace formula, see Zagier [**Zag 79**], p. 316, formula (2.12) and Szmidt [**Szm 83**]. See also [**JoL 04**].

4.7. Onward

Finally, we make further comments, along the same lines as in the Introduction (see also [**JoL 01a**]). The spectral expansion for the heat kernel as above can be viewed as a theta relation. For compact quotients, after taking the manifold trace (integrating over the diagonal), this was already observed by Gangolli [**Gan 68**]. In the present case, the finite sum over parabolics gives an additional term, which fits the pattern from [**JoL 94**], Chapter V (reproduced in [**La-Jo 01**]).

Classically, starting with Riemann, one takes a Mellin transform of a theta relation to get a zeta function with functional equation. In the above mentioned chapter, we showed how to get something which deserves to be called a zeta function by taking the Gauss transform, with a suitable regularization. Specifically, the Gauss transform of a function f on $(0, \infty)$ is more or less defined by the integral transform

$$\mathrm{Gauss}(f)(z) = 2z \int_0^\infty e^{-z^2 t} f(t)dt.$$

Usually one deals with functions which decrease rapidly at infinity, but there remains a singularity at the origin which needs to be regularized, as explained in [**JoL 94**] Chapter V. Thus the Gauss transform is essentially a Laplace transform with the change of variables $z \mapsto z^2$, together with the regularization procedure, which we recall briefly for the reader's convenience.

We suppose there is some positive integer N such that $f(t) = O(1/t^N)$ for $t \to 0$. We then define

$$\mathrm{Gauss}^{(N)}(f)(z) = 2z \int_0^\infty e^{-z^2 t} f(t) t^{N+1} \frac{dt}{t}.$$

As usual, multiplying an integrand by a power of the variable corresponds to taking derivatives outside the integral defining the function. Thus with the differential operator $\mathcal{D}_z = -(1/2z)d/dz$, $\mathrm{Gauss}^{(N)}(f)$ is the iterated derivative $\mathcal{D}_z^N$ of a function which is well defined up to a polynomial. One then has to normalize this polynomial in the manner explained in [**JoL 94**]. Roughly speaking, the polynomial is determined so that an analogue of the classical **Lerch formula** for the gamma function is satisfied in the more general setting. Cf. [**JoL 93**]. In the situation of this reference, we consider more generally the Laplace-Mellin transform

$$\mathbf{LM}f(s, z) = \int_0^\infty e^{-zt} f(t) t^s \frac{dt}{t}.$$

There is an appropriate analytic continuation, and there is a meromorphic function D (called a **regularized determinant**) such that letting $\mathrm{CT}_{s=1}$ denote the constant term at $s = 1$, and $\mathrm{CT}_{s=0}$ the constant term at $s = 0$, we have

$$D'/D(z) = \mathrm{CT}_{s=1}\mathbf{LM}\theta(s,z) = -\partial_z \mathrm{CT}_{s=0}\mathbf{LM}\theta(s,z).$$

This is a fairly general formulation of the Lerch formula. The function D determines the appropriate normalization of the Gauss transform at $s = N + 1$.

The Gauss transform leads further than the Mellin transform in certain directions. In particular, on $\mathrm{SL}_2(\mathbf{R})$, what one gets (after taking the regularized manifold trace) is the logarithmic derivative of the Selberg zeta function, up to terms representing lower level phenomena, which in this simplest case are gamma and Riemann-Dedekind zeta fudge terms. In any case, the Gauss transform yields functions which deserve to be called zeta functions with an additive formalism (functional equation). If the residues are integers, then this Gauss transform is the logarithmic derivative of a function with a multiplicative functional equation, having essentially the analytic structure of a zeta function of classical type but without necessarily having an Euler product.

Integrals over vertical lines (the one-variable case of the integrals we meet in **EFEX 1**) are analyzed in the above mentioned Chapter V, where they are called E-transforms. Part of our general program is to extend that chapter to the situation in higher rank. Be it noted, however, that the theta relations we made explicit in Section 6 is still in one variable t, and therefore so is the Gauss transform in one variable s. We shall describe elsewhere how to extend the theory to several variables. In effect, here, the one variable comes from the operator $\exp(-t\Delta)$. To get a function in several variables, one has to use other differential operators from the algebra of invariant differential operators. The one variable case is sufficient already to get a ladder with the Riemann zeta at the bottom.

In this way, the theory of explicit formulas for regularized series merges with the heated harmonic analysis on reductive Lie groups. This considerably broader context was the principal motivation for the axiomatization that we started, to make the entire set-up (regularized products or series and explicit formulas) applicable simultaneously to the cases of classical analytic number theory and the geometric cases which arise from groups like SL_n (reductive groups). Note that at levels higher than $n = 2$, the main contributions to the fudge terms arise from the parabolics, which are like the original group SL_n but of lower dimension. Thus the fudge terms (not factors because we are carrying on in an additive setting) correspond to zeta functions associated with lower steps in the ladder, in the present case the SL_n- ladder. Other ladders will ultimately include ladders arising from algebraic and differential geometry, such as ladders of moduli spaces of various types. For instance, one can already see the Siegel modular ladder corresponding to the groups Sp_{2g} (associated to abelian varieties of dimension g), the ladder of moduli spaces for $K3$-surfaces corresponding to the group $\mathrm{SO}_0(2, 19)$, and Calabi-Yau manifolds with their more complicated moduli structure, the moduli ladder of forms of higher degree as in a paper of Jordan [**Jor 1880**], etc. On the geometric side, the moduli spaces can be compactified by spaces of similar types, but lower dimension. To each stratum which is of type $\Gamma_n \backslash \mathbf{X}_n = \Gamma_n \backslash G_n / K_n$ one assigns a zeta function obtained by the above procedure:

- Start with the heat kernel on $\mathbf{X}_n$
- Periodize by Γ_n to $\Gamma_n\backslash\mathbf{X}_n$.
- Expand the periodized heat kernel in an eigenfunction expansion.
- Regularize and integrate over $\Gamma_n\backslash G_n$, thereby getting a theta inversion formula.
- Apply the Gauss transform.

In the geometric ladder, a space at a given level is compactified by the spaces of lower level. This geometric compactification is reflected in a process of completion of the associated zeta functions. Indeed, the fudge terms of the functional equation for the zeta function associated with a given stratum are mostly the zeta functions associated with the lower strata (up to terms which essentially measure some sort of singularities). Thus the geometric ladder and the ladder of zeta functions reflect each other thereby interlocking the theory of spaces coming from algebraic and differential geometry with analysis and a framework whose origins to a large extent stem from analytic number theory. Ultimately, how the algebraic geometry or differential geometry of the strata is reflected in the analytic-algebraic behavior of the zeta functions is of interest for its own sake. On the other hand, for some purposes, and in any case as a necessary preliminary for everything else, the purely analytic aspects have to be systematically available, starting with the eigenfunction expansion. No matter what, the theory of regularized products or series and heated harmonic analysis merge further with algebraic geometry and differential geometry, with no end in sight.

Appendix: The Heat Kernel

We shall here summarize briefly the definition and basic properties of the heat kernel. We don't know a suitable single reference in book form that fits our purposes, and a monograph needs to be written. In the meantime, we refer readers to Yosida [**Yos 53**], Ito [**Ito 54**], Dodziuk [**Dod 83**] and Chavel [**Cha 84**]. Note however that the latter slides over a proof of positivity from weaker conditions, a matter which is not trivial and depends in general on the maximum principle for solutions of parabolic equations. On our G/K with G complex, the positivity is seen directly from the Gangolli formula. On a real group, it follows because of the Flensted-Jensen transform. Here we give a few facts adjusted for our immediate purposes. From our point of view, we have the existence via Chapter XII of [**JoL 01a**], so we can go directly to the basic properties.

The above mentioned paper of Dodziuk is followed by Chavel "almost verbatim" and is the best reference we know for the non-compact case, which is precisely the case of interest to us. It relies on some substantial background of differential geometry, and the main tool is the maximum principle. We summarize the results of his paper. For our purposes, we take a stronger definition of the heat kernel than he does.

This appendix is not a substitute for a systematic monograph on the heat kernel, but we hope it will help the reader in tracing complete references to the literature for the facts we need.

A.1. Dodziuk's uniqueness theorem

Let X be a Riemannian manifold., Let Δ be the Laplacian, with sign convention that on $\mathbf{R}$, $\Delta = -(d/dx)^2$. We let the **heat operator** be

$$\mathbf{H} = \Delta + \frac{\partial}{\partial t},$$

acting on functions of two variables (t, x) with t in some interval of positive numbers, and $x \in X$. We may use classical notation $\mathbf{H} = \mathbf{H}_{t,x}$ and $\Delta = \Delta_x$ to indicate the variables under consideration. Let J be an open interval. A **solution of the heat equation** on $J \times X$ is a function $f : J \times X \to \mathbf{R}$ such that

$$\mathbf{H}f = 0.$$

The derivatives must make sense, so we assume in addition that f is C^2 in x and C^1 in t for $t \in J$. Usually the interval is an open interval $(0, T)$ with $T > 0$ or $T = \infty$, that is, $J = \mathbf{R}_{>0}$.

Suppose $J = (0, T)$. We are interested in the two cases when f admits a continuous extension to the half closed interval $[0, T)$, and when it does not. Both cases arise in a natural fashion. Thus we define a function

$$f : [0, T) \times X \to \mathbf{R}$$

to be an **initially complete solution of the heat equation** if it is continuous, and its restriction to $(0, T) \times X$ is C^2 in x, C^1 in t, and a solution of the heat equation. We then call the function

$$x \mapsto f(0, x) = f_0(x)$$

the **initial condition**. A function $f : (0, T) \times X \to \mathbf{R}$ will be said to **have an initial condition** if it can be extended to a continuous function on $[0, T) \times \mathbf{X}$ which is an initially complete solution of the heat equation.

Examples. On $(0, \infty) \times \mathbf{R} = \mathbf{R}_{>0} \times \mathbf{R}$, the function called the **heat gaussian**

$$\mathbf{g}(t, x) = \mathbf{g}_t(x) = \frac{1}{(4\pi t)^{1/2}} e^{-x^2/4t}$$

is a solution of the heat equation. It is not initially complete, i.e. it has no continuous extension to $t = 0$. It is bounded in x for each value of t. It will provide an example for a fundamental solution of the heat equation to be discussed shortly.

For another example, let

$$f(t, x) = -\partial_t \mathbf{g}(t, x) = \left\{ \begin{array}{ll} \dfrac{x}{(4\pi t)^{3/2}} e^{-x^2/4t} & \text{if } t \neq 0 \\ 0 & \text{if } t = 0 \end{array} \right\}.$$

Then f satisfies the heat equation for $t > 0$. However, f is not continuous at $(0, 0)$, and is also not bounded, e.g. along the curve $x = t^{\frac{1}{2}}$ for $t \to 0$. Thus f is not initially complete. Also, f is not positive.

The uniqueness theorem will be phrased under a differential geometric condition of Ricci curvature. This notion depends on the Riemann tensor R, and the literature is split about its sign convention. We pick here the convention opposite to that of Dodziuk, namely for vector fields ξ, η (on an open set),

$$R(\xi, \eta) = D_\xi D_\eta - D_\eta D_\xi - D_{[\xi, \eta]}.$$

Then the **Ricci R-tensor** is the trace of the corresponding endomorphism of the tangent bundle, that is for any orthonormal frame $\xi_1, \ldots, \xi_n$,

$$\mathrm{Ric}_R(\xi, \eta) = \sum R(\xi, \xi_j, \eta, \xi_j).$$

We then obtain a quadratic form $\xi \mapsto \mathrm{Ric}_R(\xi, \xi)$ on the tangent bundle. We can restrict this form to the unit sphere bundle. The **Ricci curvature** $\mathrm{cur}_{\mathrm{Ric}}$ is defined to be minus the retriction of this quadratic form to the unit sphere bundle, that is $-\mathrm{Ric}_R(\xi, \xi)$ for a unit vector field ξ. Thus the Ricci curvature is defined on unit vectors in the tangent bundle. Bounds from above for our Ricci form are equivalent to bounds from below for the Ricci curvature. Note that our G/K has negative curvature, so positive Ricci form.

THEOREM 1.1 (DODZIUK [**Dod 83**]). *Let X be a complete Riemannian manifold whose Ricci curvature is bounded from below. Then a bounded initially complete solution of the heat equation is uniquely determined by its initial condition.*

Dodziuk's hypotheses in Theorem 1.1 are very convenient for applications. The curvature condition is satisfied in the examples we have in mind for the following reasons. We are interested in Riemannian manifolds which have the following property, proved for general symmetric spaces in [**Bor 63**].

There exists a discrete subgroup Γ_c of Riemannian automorphisms acting freely on X such that the quotient manifold $\Gamma_c \backslash X$ is compact.

Margulis told us that on SL_n one can use the units of reduced norm 1 in a division algebra over $\mathbf{Q}$ splitting over the reals to construct Γ_c. Thus $X \to \Gamma_c \backslash X$ is a covering of a compact manifold. Local invariants from differential geometry, in particular Ricci curvature, are continuous on X or the sphere bundle, invariant under Riemannian automorphisms, and therefore continuous on the quotient space, hence bounded. It's that simple.

Given a discrete subgroup of the type we have been working with, the quotient $\Gamma \backslash \mathbf{X}$ is of course not compact, and besides, the group does not act freely. As to the latter, it is a theorem that given a discrete subgroup Γ of arithmetic type, there exists a subgroup Γ_0 of finite index in Γ such that Γ_0 acts freely on the space $\mathbf{X} = G/K$. See Borel [**Bor 69**], 17.1 and 17.4. Then $\Gamma_0 \backslash \mathbf{X}$ is a "ramified " covering of $\Gamma \backslash \mathbf{X}$, and one may simply view $C^\infty(\Gamma \backslash \mathbf{X})$ as the space of C^∞ functions on $\Gamma_0 \backslash \mathbf{X}$ which are invariant under Γ. The space $\Gamma_0 \backslash \mathbf{X}$ is a manifold, to which Dodziuk's theorem can be applied as stated.

As to the non-compactness of $\Gamma \backslash \mathbf{X}$, it is an area of interest for its own sake to compare $\Gamma \backslash \mathbf{X}$ with a compact manifold $\Gamma_c \backslash \mathbf{X}$ using another co-compact discrete subgroup Γ_c. Already on Riemann surfaces when $\mathbf{X} = \mathbf{h}_2$ is the upper half plane, one meets situations of interest. The non-compact quotient space has a compactification whose fundamental group π_1 is isomorphic to a group Γ_c of metric automorphisms of $\mathbf{h}_2$, with compact quotient. The comparison of $\Gamma \backslash \mathbf{h}_2$ with $\Gamma_c \backslash \mathbf{h}_2$ can be made from many points of view, analytic, geometric and number theoretic.

A.2. The fundamental solution and the heat kernel

There are roughly two ways to define the heat kernel. One way is via so-called "weak" properties (concerning the effect on a space of test functions), and the other is intrinsic to the kernel function itself. We discuss both, starting with the "weak" definition. We let:

$\mathrm{BC}(X)$ = space of bounded continuous functions on X.

Dodziuk defines a **fundamental solution** of the heat equation on X to be a function

$$K : (0, \infty) \times X \times X \to \mathbf{R}$$

which is continuous, such that for each (t, x) the function $y \mapsto K(t, x, y)$ is in $L^1(\mu)$ ($\mu =$ Riemannian measure), and for each $\varphi \in \mathrm{BC}(X)$ the function

$$f(t,x) = \begin{cases} (K * \varphi)(t,x) & \text{for } t > 0 \\ \varphi(x) & \text{for } t = 0 \end{cases}$$

is an initially complete solution of the heat equation.

By Theorem 1.1, such a fundamental solution is unique if X is complete and has Ricci curvature bounded from below. To avoid technical complications both mathematical and linguistic, *we assume that our manifold X satisfies these two conditions, completeness and Ricci curvature bounded from below.*

Dodziuk proves the existence of a fundamental solution via an exhaustion method for expanding submanifolds with boundary.

In the Dodziuk definition, the fundamental solution is not required to satisfy the heat equation, and only the weak limiting property

$$\lim_{t \to 0} (K * \varphi)(t, x) = \varphi(x)$$

for bounded continuous functions φ appears. Subsequently, Dodziuk does prove that the fundamental solution is C^∞ on $(0, \infty) \times X \times X$, satisfies the heat equation, and satisfies other properties which we list systematically in §3.

Note that the fundamental solution is not initially complete. It does not have an initial condition as in §1.

For our purposes, we prefer to deal with a definition which does not a priori involve test functions, but involves only properties of the kernel function itself. We do this as follows.

Let X be a metric space and μ a positive Borel measure on X. By a **Dirac family** on X, we mean a family $\{K_t\}$ indexed by $t \in \mathbf{R}_{>0}$, of continuous functions on $X \times X$, satisfying the following three conditions:

DIR 1. For all t, we have $K_t \geqq 0$ (semipositivity).

DIR 2. For all $x \in X$, we have the probabilistic condition

$$\int_X K_t(x, y) d\mu(y) = 1.$$

DIR 3. Let $d(x, y)$ be the distance function. Given $x \in X$ and $\delta > 0$, we have

$$\lim_{t \to 0} \int_{d(x,y) \geqq \delta} K_t(x, y) d\mu(y) = 0.$$

We recall that if φ is a function on X then the **convolution** $K_t * \varphi$ is defined by the integral

$$(K_t * \varphi)(x) = \int_X K_t(x, y) \varphi(y) d\mu(y).$$

We are indebted to Sattinger for pointing out that the first manifestation of the following theorem on the real line dates back to Weierstrass [**Wei 1885**].

THEOREM 2.1. *Let $\{K_t\}$ be a Dirac family. Let φ be bounded measurable on X. Then for every x where φ is continuous, we have the pointwise limit*

$$\lim_{t\to 0}(K_t * \varphi)(x) = \varphi(x),$$

uniformly on every set where φ is continuous, and where the limit in **DIR 3** *is uniform for given δ.*

PROOF. We have

$$(K_t * \varphi)(x) - \varphi(x) = \int_X K_t(x,y)[\varphi(y) - \varphi(x)]d\mu(y).$$

We estimate the absolute value on the left, namely

$$\begin{aligned}|(K_t * \varphi)(x) - \varphi(x)| &\leqq \int_X K_t(x,y)|\varphi(y) - \varphi(x)|d\mu(y)\\ &\leqq \int_{d(x,y)<\delta} + \int_{d(x,y)\geqq\delta} K_t(x,y)|\varphi(y) - \varphi(x)|d\mu(y).\end{aligned}$$

Given ε, we pick δ so that for $d(x,y) < \delta$ we have $|\varphi(y) - \varphi(x)| < \varepsilon$. Then by **DIR 2**, the first integral is bounded by ε. The second integral is bounded by $2\|\varphi\|_\infty$ (sup norm) times the integral of K_t, which tends to 0 as $t \to \infty$ by **DIR 3**. This concludes the proof.

In light of the theorem, a Dirac family is sometimes called an **approximation of the identity**.

We define a **heat kernel** on X to be a C^∞ function

$$\mathbf{K} : (0,\infty) \times X \times X \to \mathbf{R}$$

satisfying the following two conditions.

HK 1. $\mathbf{K}$ satisfies the heat equation in the first two variables (t,x).

HK 2. $\{\mathbf{K}_t\}$ is a Dirac family.

In the context of semi-simple Lie groups, an existence proof comes from the spherical inversion theory, whereby the heat kernel is defined as the inverse spherical transform of a suitably normalized Gaussian on euclidean space, as in [**Gan 68**]. To summarize, let this Gaussian on $\mathbf{i}\mathfrak{a}^\vee$ be defined by

$$E_t(\zeta) = \exp\left((\zeta^2 - \rho^2)t\right),$$

where $\zeta^2 = \langle\zeta,\zeta\rangle$, ζ is restricted to the imaginary axis $\zeta = \mathbf{i}\lambda$, $\lambda \in \mathfrak{a}^\vee$, and we recall that $\zeta^2 - \rho^2 = \mathrm{ev}(\omega, \varphi_\zeta)$ is the eigenvalue of the Casimir ω on the spherical function φ_ζ. Then $\mathbf{g}_t$ is the spherical inverse transform of E_t, namely

$$\mathbf{S}\mathbf{g}_t = E_t \quad \text{or also} \quad \mathbf{g}_t = \mathbf{S}^{-1}E_t.$$

This existence proof yields the Gangolli formula in the complex case, reproduced in Chapter 2, §2. Existence for real groups then follows from the Flensted-Jensen transform [**Fle 78**]. Cf. [**JoL 01a**], Chapter XII, §6.

For a more general point of view, existence and uniqueness can be proved from the following theorem.

THEOREM 2.2. *Under the general conditions that X is complete and has Ricci curvature bounded from below, a heat kernel is equal to the fundamental solution.*

For a proof, Grigoryan has told us two more structural ways. One way depends on developing the theory of the heat kernel from an L^2-functional analysis point of view. Another way consists in following the Dodziuk exhaustion construction, building in from the start estimates required to give the desired implication. Grigoryan told us that a proof can thus be extracted from [**Dod 83**] and also [**ChY 81**].

On the other hand, there is also a direct approach if one can show that **HK 1** implies the heat equation for $\mathbf{K}_t * \varphi$ by using DUTIS (Differentiating Under The Integral Sign), as follows.

LEMMA 2.3. *Suppose in addition to* **HK 1**, **HK 2**, *that* $\mathbf{K}$ *satisfies the uniform estimate as follows.*

HK 3. *For each point of $(0, \infty) \times X$ there is a chart U with coordinates $(t, x_1, \dots, x_N)$ such that there is a bounded positive function $g \in L^1(\mu)$ on X having the following property. For every monomial D of partial derivatives of the form ∂_t, or $1, \partial_i, \partial_i \partial_j$ in the X-variables, we have for all $(t, x) \in U, y \in X$ the estimate*

$$|(D\mathbf{K})(t, x, y)| \leqq g(y).$$

Then a heat kernel satisfying **HK 3** *is a fundamental solution in the Dodziuk sense.*

The estimate of **HK 3** is adjusted to conditions which can be verified directly on $\mathbf{K}$ in important cases. For example, with the Gangolli formula, one has a structural way of verifying the conditions via the theory of spherical functions, cf. [**JoL 01a**] Chapter X, §7 and Chapter XII, §5. We then apply the following calculus lemma.

LEMMA 2.4. *Let U be an open set in a euclidean space, and let Y be a Riemannian manifold. Let*

$$f : U \times Y \to \mathbf{R}$$

be a C^∞ function. Let $x = (x_1, \dots, x_N)$ be the real coordinates in U. Suppose that there exist functions $g_0(y), g_1(y)$ $(y \in Y)$ which are positive, such that

$$|f(x, y)| \leqq g_0(y) \quad \text{and} \quad |\partial_i f(x, y)| \leqq g_1(y) \text{ for } i = 1, \dots, N$$

for all x, y and such that $g_0, g_1 \in L^1(\mu)$. Let

$$F(x) = \int_X f(x, y) d\mu(y).$$

Then F is continuous, the partials $\partial_i F$ exist, and

$$\partial_i F(x) = \int_X \partial_i f(x, y) d\mu(y).$$

The proof is standard calculus, see for instance [**Lan 83/97**], Chapter XIII, §3.

This concludes our general discussion of various definitions of the heat kernel.

Example. Let $\mathbf{g}_t$ be the function on $\mathbf{R}$ given as an example previously, that is

$$\mathbf{g}_t(x) = \frac{1}{(4\pi t)^{1/2}} e^{-x^2/4t}.$$

Let $\mathbf{K}(t, x, y) = \mathbf{g}_t(x - y) = \mathbf{g}_t(y - x)$. Then $\mathbf{K}$ is the fundamental solution and the heat kernel on $\mathbf{R}$. On G/K, See Chapter 2, §2.

A.3. Properties of the heat kernel

We now list further properties of the heat kernel, using our definition.

THEOREM 3.1. *Let X be a complete Riemannian manifold with Ricci curvature bounded from below. Then the heat kernel satisfies the following additional properties.*

Positivity. *It is strictly positive, i.e.* $\mathbf{K}(t, x, y) > 0$ *for all* t, x, y.
Uniqueness. *It is uniquely determined, and is the minimal positive fundamental solution.*
Symmetry. *It is symmetric, that is* $\mathbf{K}_t(x, y) = \mathbf{K}_t(y, x)$ *for all* t, x, y.
Semigroup. *For* $t, s > 0$, *we have*

$$\mathbf{K}_t * \mathbf{K}_s = \mathbf{K}_{t+s},$$

or written in full, for $x, y \in X$,

$$\int_X \mathbf{K}(t, x, z)\mathbf{K}(s, z, y) d\mu(z) = \mathbf{K}(t + s, x, y).$$

Adjointness. *For* φ, ψ *bounded measurable, and* $\psi \in L^1(\mu)$, *we have*

$$\langle \mathbf{K}_t * \varphi, \psi \rangle = \langle \varphi, \mathbf{K}_t * \psi \rangle.$$

These properties are proved by Dodziuk for the fundamental solution. Note that the uniqueness comes directly from Theorem 1.1, and the last property is just an application of Fubini's theorem. On the other hand, except for the uniqueness, in our context the properties also come out of the theory of spherical functions, cf. [**JoL 01a**], Chapter X, §7 and Chapter XII, §5.

The adjointness property is immediate from Fubini's theorem. As a consequence, one shows that the convolution operator $\varphi \mapsto \mathbf{K}_t * \varphi$ is semipositive, by using the semigroup property, namely $\mathbf{K}_t = \mathbf{K}_{t/2} * \mathbf{K}_{t/2}$, and

$$\langle \mathbf{K}_t * \varphi, \varphi \rangle = \langle \mathbf{K}_{t/2} * \mathbf{K}_{t/2} * \varphi, \varphi \rangle = \langle \mathbf{K}_{t/2} * \varphi, \mathbf{K}_{t/2} * \varphi \rangle \geqq 0.$$

However, for the positivity or even semipositivity $\mathbf{K}(t,x,y) \geqq 0$, or $\mathbf{K}(t,x,y) > 0$, one needs the maximum principle.

A.4. Compact manifolds

One method for proofs of the heat kernel's properties is via Duhamel's formula as in [**Dod 83**] and [**Cha 84**]. On compact manifolds, no convergence properties intervene with the formal arguments, so we carry out the arguments in this case to show how the method works. From here on:

Let X be a compact Riemannian manifold.

THEOREM 4.1. *Let $t_1 > 0$. Let φ, ψ be C^∞ functions on $(0, t_1) \times X$. Let*

$$0 < a \leqq b < t < t_1.$$

Then for $z \in X$, we have **Duhamel's formula**

$$\int_X [\varphi(t-b,z)\psi(b,z) - \varphi(t-a,z)\psi(a,z)]d\mu(z)$$

$$= \int_a^b d\tau \int_X [\varphi(t-\tau,z)\mathbf{H}\psi(\tau,z) - \mathbf{H}\varphi(t-\tau,z)\psi(\tau,z)]d\mu(z).$$

PROOF. Let $\varphi_t(\tau,z) = \varphi(t-\tau,z)$. Using the definition $\mathbf{H} = \Delta + \partial_1$, we get:

$$\begin{aligned}\varphi_t\mathbf{H}\psi - \psi\mathbf{H}\varphi_t &= \varphi_t\Delta\psi - \psi\Delta\varphi_t + \varphi_t\partial_1\psi + \psi\partial_1\varphi_t \\ &= \varphi_t\Delta\psi - \psi\Delta\varphi_t + \partial_1(\varphi_t\psi).\end{aligned}$$

Green's theorem, applied to the case when X has no boundary, tells us that

$$\int_X (\varphi_t\Delta\psi - \psi\Delta\varphi_t)d\mu = 0.$$

See for instance [**Lan 99**], Chapter XVIII, Theorem 3.4. Then the double integral on the right of Duhamel's formula is

$$\begin{aligned}\int_a^b d\tau \int_X [\ldots]d\mu(z) &= \int_a^b \int_X \partial_1(\varphi_t\psi)(r,z)d\mu(z)dr \\ &= \int_X (\varphi_t\psi)(r,z)\Big|_{r=a}^{r=b} d\mu(z)\end{aligned}$$

which proves the Duhamel formula.

REMARK. In the non-compact case, one has to take into account either boundary terms arising from truncation and an exhaustion function, and sufficiently rapid decay of these boundary terms at infinity. The boundary terms involve first and second derivatives, arising from Δ and $\partial/\partial t$.

THEOREM 4.2. *A heat kernel is uniquely determined, and is symmetric, that is*

$$\mathbf{K}_t(x,y) = \mathbf{K}_t(y,x) \quad \textit{for all } x, y \in X \textit{ and } t > 0.$$

PROOF. Let F be another heat kernel, and fix $x, y \in X$. Let

$$\varphi(t,z) = \mathbf{K}(t,x,z) \text{ and } \psi(t,z) = F(t,y,z).$$

Then $\mathbf{H}\varphi = \mathbf{H}\psi = 0$. Duhamel's formula gives

$$\int_X \mathbf{K}(t-b,x,z)f(b,y,z)d\mu(z) = \int_X \mathbf{K}(t-a,x,z)F(a,y,z)d\mu(z). \tag{1}$$

Let $b \to t$. We claim that the limit of the left side of (1) is equal to

$$\lim_{u\to 0}\int_X \mathbf{K}(u,x,z)F(t,y,z)d\mu(z) = F(t,y,x). \tag{2}$$

Proof. We write first the left side of (1) in the form

$$\int_X \mathbf{K}(t-b,x,z)[F(b,y,z) - F(t,y,z)]d\mu(z) + \int_X \mathbf{K}(t-b,x,z)F(t,y,x)d\mu(x). \tag{3}$$

We estimate the first term of (3). Given ε we can find a δ-neighborhood of t such that for b in this neighborhood, we have for all $z \in X$,

$$|F(b,y,z) - F(t,y,z)| < \varepsilon.$$

Hence the first term of (3) is bounded by

$$\varepsilon \int_X \mathbf{K}(t-b,x,z)d\mu(z) = \varepsilon.$$

Letting $b \to t$ we conclude that the first term in (3) approaches 0 as $b \to t$. Hence the limit of the left side of (1) is the same as

$$\lim_{u\to 0}\int_X \mathbf{K}(u,x,z)F(t,y,z)d\mu(z).$$

Now we can apply the Dirac property to conclude the proof of (2).

By a permutation of the alphabet, the limit of the right side of (1) is equal to $\mathbf{K}(t,x,y)$. Taking $F = \mathbf{K}$ proves the symmetry. Then for arbitrary F, we get the uniqueness. This concludes the proof of Theorem 4.2.

One also has a uniqueness statement of solutions of the heat equation with given initial conditions. We assume that the heat kernel exists.

THEOREM 4.3. *Let f be a continuous function on X. Let F be a C^∞ function on $\mathbf{R}_{>0} \times X$. Suppose F satisfies the heat equation and has an initial condition, or if you wish,*

$$\lim_{t\to 0} F(t,x) = f(x) \text{ uniformly for } x \in X.$$

Then $F = \mathbf{K} * f$, that is for all $x \in X$,

$$F(t,x) = (\mathbf{K}_t * f)(x).$$

PROOF. Given $x \in X$, let $\varphi = F$ and $\psi(\tau, z) = \mathbf{K}(\tau, x, z)$ in Duhamel's formula. By hypothesis that $\mathbf{H}F = \mathbf{H}\mathbf{K} = 0$, letting $0 < a \leqq b < t < t_1$ as in Duhamel, we obtain

$$\int_X F(t-b,z)\mathbf{K}(b,x,z)d\mu(z) = \int_X F(t-a,z)\mathbf{K}(a,x,z)d\mu(z).$$

Let $b \uparrow t$ so $t - b \to 0$. Then $F(t-b,z) \to f(z)$, and the left side approaches $(\mathbf{K} * f)(t,x)$. Let $a \downarrow 0$. Then the right side approaches

$$\lim_{a\to 0} \int_X f(t,z)\mathbf{K}(a,x,z)d\mu(z) = F(t,x).$$

This concludes the proof.

THEOREM 4.4. *The heat kernel satisfies the one-parameter semigroup property, that is for $s, t > 0$,*

$$\mathbf{K}_t * \mathbf{K}_s = \mathbf{K}_{t+s}.$$

As usual by definition,

$$\mathbf{K}_t * \mathbf{K}_s(x,y) = \int_X \mathbf{K}_t(x,z)\mathbf{K}_s(y,z)d\mu(z).$$

PROOF. For x, y fixed in X, let

$$\varphi(\tau,z) = \mathbf{K}(\tau+s,z,y) \quad \text{and} \quad \psi(\tau,z) = \mathbf{K}(\tau,x,z).$$

Then $\mathbf{H}\varphi = \mathbf{H}\psi = 0$. Hence Duhamel's formula gives

$$\int_X \mathbf{K}(t+s-b,y,z)\mathbf{K}(b,x,z)d\mu(z) = \int_X \mathbf{K}(t+s-a,y,z)\mathbf{K}(a,x,z)d\mu(z).$$

Let $b \uparrow t$. Then the left side approaches

$$\int_X \mathbf{K}(s,y,z)\mathbf{K}(t,x,z)d\mu(z).$$

Let $a \downarrow 0$. We get from the Dirac property that the right side approaches $\mathbf{K}(t + s, y, x)$, which concludes the proof.

On a non-compact manifold, one can follow a similar route, but one has to take into account growth conditions and convergence conditions. The above method with

Duhamel's formula works, but the boundary components have to be estimated, and need to tend to 0 in an exhaustion of the manifold by submanifolds with boundary. Cf. the reference given in the introduction to this appendix.

Bibliography

[Bor 63] A. Borel, *Compact Clifford Klein forms of symmetric spaces*, Topology **2** (1963), 111-122.

[Bor 69] A. Borel, *Introduction aux groupes arithmétiques*, Hermann, Paris, 1969.

[Bor 91] A. Borel, *Linear algebraic grops, Second enlarged edition*, Springer Verlag, New York, 1991.

[Bor 97] A. Borel, *Automorphic forms on* $\mathrm{SL}_2(\mathbf{R})$, Cambridge University Press, Cambridge, UK, 1997.

[BoG 83] A. Borel and H. Garland, *Laplacian and the discrete spectrum of an arithmetic group*, Amer. J. Math **105** (1983), 309-355.

[BoT 65] A. Borel and J. Tits, *Groupes réductifs*, Publ. Math. IHES **27** (1965), 55-150.

[Cha 84] I. Chavel, *Eigenvalues in Riemannian Geometry*, Academic Press, New York, 1984.

[ChY 81] J. Cheeger and S.-T. Yau, *A lower bound for the heat kernel*, Comm. pure. appl. math. **34** (1981), 465-480.

[Che 56/58] C. Chevalley, *Séminaire sur la classification des groupes de Lie algébriques, I, II*, Paris, 1956-58.

[Dod 83] J. Dodziuk, *Maximum principle for parabolic inequalities and heat flow on open manifolds*, Indiana Univ. Math. J. **32** (1983), 703-716.

[Fle 78] M. Flensted-Jensen, *Spherical functions on a real semisimple Lie group: A method of reduction to the complex case*, J. Funct. Analysis **30** (1978), 106-146.

[Gan 68] R. Gangolli, *Asymptotic behavior of spectra of compact quotients of certain symmetric spaces*, Acta Math. **121** (1968), 151-192.

[GaV 88] R. Gangolli and V. Varadarajan, *Harmonic Analysis of Spherical Functions on Real Reduction Groups*, Springer Verlag, New York, 1988.

[GPS 63] I. Gelfand and I. Piatetski-Shapiro, *Automorphic functions and representation theory*, Trudy Moscow Math. Obsc. **12** (1963), 389-412.

[God 57] R. Godement, *Introduction aux travaux de Selberg*, Séminaire Bourbaki, 1957.

[God 66] R. Godement, *The spectral decomposition of cusp forms*, Proc. Sympos. Appl. Math., vol. 9, Amer. Math. Soc., Providence, RI, 1966, pp. 225-234.

[God 67] R. Godement, *Introduction aux travaux de Langlands*, Séminaire Bourbaki, 1967.

[Gre 88] D. Grenier, *Fundamental domains for the general linear group*, Pacific J. Math. **132** (1988), 293-317.

[Gre 93] D. Grenier, *On the shape of fundamental domains in* $\mathrm{GL}(n, \mathbf{R})/\mathrm{O}(n)$, Pacific J. Math. **150** (1993), 53-66.

[Har 59] Harish-Chandra, *Automorphic forms on a semisimple Lie group*, Proc. Natl. Acad. Sci. USA **45** (1959), 570-573.

[Har 65] Harish-Chandra, *Invariant eigendistributions on a semisimple Lie group*, Publ. IHES **27** (1965), 5-54.

[Har 66] Harish-Chandra, *Discrete series for semisimple Lie groups, II*, Acta Math. **116** (1966), 1-111.

[Har 68] Harish-Chandra, *Automorphic forms on semisimple Lie groups*, Springer Lecture Notes in Mathematics 62, New York, 1968.

[Har 75] Harish-Chandra, *Harmonic analysis on real reductive groups I. The theory of the constant term*, J. Funct. Analysis **19** (1975), 104-204.

[Hej 83] D. Hejhal, *The Selberg trace formula for* $\mathrm{PSL}(2, \mathbf{R})$, Springer Lecture Notes in Mathematics 1001, New York, 1983.

[Hel 84] S. Helgason, *Groups and geometric analysis*, Academic Press, New York, 1984.

[Hel 94] S. Helgason, *Geometric analysis on symmetric spaces*, Math Survey & Monographs 39, AMS,, Providence, R. I., 1994.

[Ito 54] S. Ito, *The fundamental solution of the parabolic equation in a differentiable manifold, II*, Osaka Math. J. **6** (1954).

[Jor 1880] C. Jordan, *Mémoire sur l'équivalence des formes*, J. École Polytechnique **XLVIII** (1880), 112-150.

[JoL 93] J. Jorgenson and S. Lang, *Basic analysis of regularized series and products*, Springer Lecture Notes in Mathematics 1564, New York, 1993.

[JoL 94] J. Jorgenson and S. Lang, *Explicit formulas for regularized products and series*, Springer Lecture Notes in Mathematics 1593, New York, 1994, pp. 1-134.

[JoL 01a] J. Jorgenson and S. Lang, *Spherical inversion on* $\mathrm{SL}_n(\mathbf{R})$, Springer Monographs in Mathematics, New York, 2001.

[JoL 01b] J. Jorgenson and S. Lang, $\mathrm{Pos}_n(\mathbf{R})$ *and Eisenstein series*, Springer Lecture Notes in Mathematics 1868, New York, 2005.

[JoL 01c] J. Jorgenson and S. Lang, *The ubiquitous heat kernel*, Mathematics unlimited: 2001 and beyond, Springer Verlag, New York, 2001, pp. 655-682.

[JoL 04a] J. Jorgenson and S. Lang, *A gaussian space of test functions*, Math. Nachr **278** (2005), 824-832.

[JoL 04b] J. Jorgenson and S. Lang, *The heat kernel and theta inversion on* $\mathrm{SL}_2(\mathbf{C})$, Springer Monographs in Mathematics, New York, 2008.

[JoL 04c] J. Jorgenson and S. Lang, *The heat kernel, theta inversion, and zetas on* $\Gamma\backslash G/K$,, to appear in *Lang Memorial Volume.*

[JSL 02] J. Jorgenson, S. Lang, and A. Sinton, *The heat kernel on totally geodesic embeddings of symmetric spaces*, to appear.

[Kub 74] T. Kubota, *Elementary theory of Eisenstein series*, Halsted Press, 1973.

[Lang 75/85] S. Lang, $\mathrm{SL}_2(\mathbf{R})$, Addison Wesley 1975, Springer Verlag 1985, New York.

[Lang 83/97] S. Lang, *Undergraduate analysis, Second edition*, Springer Verlag, New York, 1997.

[Lang 93b] S. Lang, *Real and functional analysis*, Springer Verlag, Graduate texts in mathematics 142, New York, 1993.

[Lang 99] S. Lang, *Fundamentals of differential geometry*, Springer Verlag, Graduate texts in mathematics 191, New York, 1999.

[LanJo 91] S. Lang, *Collected Papers vol. V (with Jay Jorgenson)*, Springer Verlag, New York, 2001.

[Llds 66] R. Langlands, *Eisenstein series*, Proc. Sympos. Appl. Math., vol. 9, Amer. Math. Soc., Providence, RI, 1966, pp. 235-252.

[Llds 76] R. Langlands, *On the functional equations satisfied by Eisenstein series*, Springer Lecture Notes in Mathematics 544, New York, 1976.

[Maa 71] H. Maass, *Siegel's modular forms and Dirichlet series*, Springer Lecture Notes in Mathematics 216, New York, 1971.

[MoW 94] C. Moeglin and J.-L. Waldspurger, *Décomposition spectrale et séries d'Eisenstein*, Birkhäuser Progress in Mathematics 113, Boston, 1994.

[Mos 55] G. Mostow, *Some new decomposition theorems for semisimple Lie groups*, Memoirs AMS 14,, Providence, R. I., 1955, pp. 31-54.

[Mue 83] W. Müller, *Spectral theory for Riemannian manifolds with cusps and a related trace formula*, Math. Nachr **111** (1983), 197-288.

[Mue 89] W. Müller, *The trace class conjecture in the theory of automorphic forms*, Annals Math. **130** (1989), 473-429.

[Mue 98] W. Müller, *The trace class conjecture in the theory of automorphic forms II*, Geom. and Funct. Analysis (GAFA) **8** (1998), 315-355.

[Mue 00] W. Müller, *On the singularities of residual intertwining operators*, Geom. and Funct. Analysis (GAFA) **10** (2000), 1118-1170.

[ProW 67] M. Protter and F. Weinberger, *Maximum principles in differential equations*, Prentice-Hall, 1967.

[Rag 72] M. Raghunathan, *Discrete subgroups of Lie groups*, Ergebinisse der Math 63, Springer-Verlag, New York, 1972.

[Roe 66] W. Roelcke, *Das Eigenwertproblem der Automorphen Formen in der Hyperbolischen Ebene I*, Math. Ann. **167** (1966), 292-337.

[Roe 67] W. Roelcke, *Das Eigenwertproblem der Automorphen Formen in der Hyperbolischen Ebene II*, Math. Ann. **168** (1967), 261-324.

[Sar 83] P. Sarnak, *The arithmetic and geometry of some hyperbolic three-manifolds*, Acta Math. **151** (1983), 253-295.

[Sel 56] A. Selberg, *Harmonic analysis and discontinuous groups*, J. Indian Math. Soc **XX** (1956), 47-87.

[Sel 62] A. Selberg, *Discontinuous groups and harmonic analysis*, Proc. Internal. Cong. Math. Stockholm (1962), 177-189.

[Spr 98] T. Springer, *Linear algebraic groups, second edition*, Birkhäuser, Boston, 1998.

[StW 71] E. Stein and G. Weiss, *Introduction to Fourier analysis on Euclidean space*, Princeton University Press, Princeton, NJ, 1971.

[Szm 83] J. Szmidt, *The Selberg trace formula for the Picard group* $\mathrm{SL}(2,\mathbf{Z}[\mathbf{i}])$, Acta Arith. **XLII** (1983), 391-424.

[Szm 87] J. Szmidt, *The Selberg trace formula and imaginary quadratic number fields*, Schriftenreihe des Sonderforschungsbereichs geometrie under analysis **52** (1987).

[Ter 85] A. Terras, *Harmonic analysis on symmetric spaces and applications I*, Springer-Verlag, New York, 1985.

[Ter 88] A. Terras, *Harmonic analysis on symmetric spaces and applications II*, Springer-Verlag, New York, 1988.

[Var 77] V. Varadarjan, *Harmonic analysis on real reductive groups*, Springer Lecture Notes in Mathematics 576, New York, 1977.

[Wei 1885] K. Weierstrass, *Über die analytische Darstellbarkeit sogenannter willkürlicher Functionen einer reelen Veränderlichen*, Sitzungsbericht Königl. Akad. Wiss. (2 and 30 July 1885), 633-639 and 789-805.

[War 72] G. Warner, *Harmonic analysis on semisimple Lie groups, Vols. I and II*, Springer Verlag, New York, 1972.

[Yos 53] K. Yosida, *On the fundamental solution of the parabolic equation in a Riemannian space*, Osaka Math. J. **5** (June 1953).

[Zag 79] D. Zagier, *Eisenstein series and the Selberg trace formula*, Automorphic Forms, Representation Theory, and Arithmetic, Tata Institute of Fundamental Research, Bombay, 1979.

Index

Editorial Information

To be published in the *Memoirs*, a paper must be correct, new, nontrivial, and significant. Further, it must be well written and of interest to a substantial number of mathematicians. Piecemeal results, such as an inconclusive step toward an unproved major theorem or a minor variation on a known result, are in general not acceptable for publication.

Papers appearing in *Memoirs* are generally at least 80 and not more than 200 published pages in length. Papers less than 80 or more than 200 published pages require the approval of the Managing Editor of the Transactions/Memoirs Editorial Board.

As of May 31, 2009, the backlog for this journal was approximately 11 volumes. This estimate is the result of dividing the number of manuscripts for this journal in the Providence office that have not yet gone to the printer on the above date by the average number of monographs per volume over the previous twelve months, reduced by the number of volumes published in four months (the time necessary for preparing a volume for the printer). (There are 6 volumes per year, each usually containing at least 4 numbers.)

A Consent to Publish and Copyright Agreement is required before a paper will be published in the *Memoirs*. After a paper is accepted for publication, the Providence office will send a Consent to Publish and Copyright Agreement to all authors of the paper. By submitting a paper to the *Memoirs*, authors certify that the results have not been submitted to nor are they under consideration for publication by another journal, conference proceedings, or similar publication.

Information for Authors

Memoirs are printed from camera copy fully prepared by the author. This means that the finished book will look exactly like the copy submitted.

Initial submission. The AMS uses Centralized Manuscript Processing for initial submissions. Authors should submit a PDF file using the Initial Manuscript Submission form found at `www.ams.org/peer-review-submission`, or send one copy of the manuscript to the following address: Centralized Manuscript Processing, MEMOIRS OF THE AMS, 201 Charles Street, Providence, RI 02904-2294 USA. If a paper copy is being forwarded to the AMS, indicate that it is for it Memoirs and include the name of the corresponding author, contact information such as email address or mailing address, and the name of an appropriate Editor to review the paper (see the list of Editors below).

The paper must contain a *descriptive title* and an *abstract* that summarizes the article in language suitable for workers in the general field (algebra, analysis, etc.). The *descriptive title* should be short, but informative; useless or vague phrases such as "some remarks about" or "concerning" should be avoided. The *abstract* should be at least one complete sentence, and at most 300 words. Included with the footnotes to the paper should be the 2000 *Mathematics Subject Classification* representing the primary and secondary subjects of the article. The classifications are accessible from `www.ams.org/msc/`. The list of classifications is also available in print starting with the 1999 annual index of *Mathematical Reviews*. The Mathematics Subject Classification footnote may be followed by a list of *key words and phrases* describing the subject matter of the article and taken from it. Journal abbreviations used in bibliographies are listed in the latest *Mathematical Reviews* annual index. The series abbreviations are also accessible from `www.ams.org/msnhtml/serials.pdf`. To help in preparing and verifying references, the AMS offers MR Lookup, a Reference Tool for Linking, at `www.ams.org/mrlookup/`.

Electronically prepared manuscripts. The AMS encourages electronically prepared manuscripts, with a strong preference for $\mathcal{AMS}$-LaTeX. To this end, the Society has prepared $\mathcal{AMS}$-LaTeX author packages for each AMS publication. Author packages include instructions for preparing electronic manuscripts, samples, and a style file that generates the particular design specifications of that publication series. Though $\mathcal{A}_{\mathcal{M}}\mathcal{S}$-LaTeX is the highly preferred format of TeX, author packages are also available in $\mathcal{A}_{\mathcal{M}}\mathcal{S}$-TeX.

Authors may retrieve an author package for *Memoirs of the AMS* from `www.ams.org/journals/memo/memoauthorpac.html` or via FTP to `ftp.ams.org` (login as `anonymous`, enter username as password, and type `cd pub/author-info`). The *AMS Author Handbook* and the *Instruction Manual* are available in PDF format from the author package link. The author package can also be obtained free of charge by sending email to `tech-support@ams.org` (Internet) or from the Publication Division, American Mathematical Society, 201 Charles St., Providence, RI 02904-2294, USA. When requesting an author package, please specify $\mathcal{A}_{\mathcal{M}}\mathcal{S}$-LaTeX or $\mathcal{A}_{\mathcal{M}}\mathcal{S}$-TeX and the publication in which your paper will appear. Please be sure to include your complete mailing address.

After acceptance. The final version of the electronic file should be sent to the Providence office (this includes any TeX source file, any graphics files, and the DVI or PostScript file) immediately after the paper has been accepted for publication.

Before sending the source file, be sure you have proofread your paper carefully. The files you send must be the EXACT files used to generate the proof copy that was accepted for publication. For all publications, authors are required to send a printed copy of their paper, which exactly matches the copy approved for publication, along with any graphics that will appear in the paper.

Accepted electronically prepared files can be submitted via the web at `www.ams.org/submit-book-journal/`, sent via FTP, or sent on CD-Rom or diskette to the Electronic Prepress Department, American Mathematical Society, 201 Charles Street, Providence, RI 02904-2294 USA. TeX source files, DVI files, and PostScript files can be transferred over the Internet by FTP to the Internet node `ftp.ams.org` (130.44.1.100). When sending a manuscript electronically via CD-Rom or diskette, please be sure to include a message identifying the paper as a Memoir.

Electronically prepared manuscripts can also be sent via email to `pub-submit@ams.org` (Internet). In order to send files via email, they must be encoded properly. (DVI files are binary and PostScript files tend to be very large.)

Electronic graphics. Comprehensive instructions on preparing graphics are available at `www.ams.org/authors/journals.html`. A few of the major requirements are given here.

Submit files for graphics as EPS (Encapsulated PostScript) files. This includes graphics originated via a graphics application as well as scanned photographs or other computer-generated images. If this is not possible, TIFF files are acceptable as long as they can be opened in Adobe Photoshop or Illustrator. No matter what method was used to produce the graphic, it is necessary to provide a paper copy to the AMS.

Authors using graphics packages for the creation of electronic art should also avoid the use of any lines thinner than 0.5 points in width. Many graphics packages allow the user to specify a "hairline" for a very thin line. Hairlines often look acceptable when proofed on a typical laser printer. However, when produced on a high-resolution laser imagesetter, hairlines become nearly invisible and will be lost entirely in the final printing process.

Screens should be set to values between 15% and 85%. Screens which fall outside of this range are too light or too dark to print correctly. Variations of screens within a graphic should be no less than 10%.

Inquiries. Any inquiries concerning a paper that has been accepted for publication should be sent to `memo-query@ams.org` or directly to the Electronic Prepress Department, American Mathematical Society, 201 Charles St., Providence, RI 02904-2294 USA.

Editors

This journal is designed particularly for long research papers, normally at least 80 pages in length, and groups of cognate papers in pure and applied mathematics. Papers intended for publication in the *Memoirs* should be addressed to one of the following editors. The AMS uses Centralized Manuscript Processing for initial submissions to AMS journals. Authors should follow instructions listed on the Initial Submission page found at `www.ams.org/memo/memosubmit.html`.

Algebra to ALEXANDER KLESHCHEV, Department of Mathematics, University of Oregon, Eugene, OR 97403-1222; email: `ams@noether.uoregon.edu`

Algebraic geometry to DAN ABRAMOVICH, Department of Mathematics, Brown University, Box 1917, Providence, RI 02912; email: `amsedit@math.brown.edu`

Algebraic geometry and its applications to MINA TEICHER, Emmy Noether Research Institute for Mathematics, Bar-Ilan University, Ramat-Gan 52900, Israel; email: `teicher@macs.biu.ac.il`

Algebraic topology to ALEJANDRO ADEM, Department of Mathematics, University of British Columbia, Room 121, 1984 Mathematics Road, Vancouver, British Columbia, Canada V6T 1Z2; email: `adem@math.ubc.ca`

Combinatorics to JOHN R. STEMBRIDGE, Department of Mathematics, University of Michigan, Ann Arbor, Michigan 48109-1109; email: `JRS@umich.edu`

Commutative and homological algebra to LUCHEZAR L. AVRAMOV, Department of Mathematics, University of Nebraska, Lincoln, NE 68588-0130; email: `avramov@math.unl.edu`

Complex analysis and harmonic analysis to ALEXANDER NAGEL, Department of Mathematics, University of Wisconsin, 480 Lincoln Drive, Madison, WI 53706-1313; email: `nagel@math.wisc.edu`

Differential geometry and global analysis to CHRIS WOODWARD, Department of Mathematics, Rutgers University, 110 Frelinghuysen Road, Piscataway, NJ 08854; email: `ctw@math.rutgers.edu`

Dynamical systems and ergodic theory and complex analysis to YUNPING JIANG, Department of Mathematics, CUNY Queens College and Graduate Center, 65-30 Kissena Blvd., Flushing, NY 11367; email: `Yunping.Jiang@qc.cuny.edu`

Functional analysis and operator algebras to DIMITRI SHLYAKHTENKO, Department of Mathematics, University of California, Los Angeles, CA 90095; email: `shlyakht@math.ucla.edu`

Geometric analysis to WILLIAM P. MINICOZZI II, Department of Mathematics, Johns Hopkins University, 3400 N. Charles St., Baltimore, MD 21218; email: `trans@math.jhu.edu`

Geometric topology to MARK FEIGHN, Math Department, Rutgers University, Newark, NJ 07102; email: `feighn@andromeda.rutgers.edu`

Harmonic analysis, representation theory, and Lie theory to ROBERT J. STANTON, Department of Mathematics, The Ohio State University, 231 West 18th Avenue, Columbus, OH 43210-1174; email: `stanton@math.ohio-state.edu`

Logic to STEFFEN LEMPP, Department of Mathematics, University of Wisconsin, 480 Lincoln Drive, Madison, Wisconsin 53706-1388; email: `lempp@math.wisc.edu`

Number theory to JONATHAN ROGAWSKI, Department of Mathematics, University of California, Los Angeles, CA 90095; email: `jonr@math.ucla.edu`

Number theory to SHANKAR SEN, Department of Mathematics, 505 Malott Hall, Cornell University, Ithaca, NY 14853; email: `ss70@cornell.edu`

Partial differential equations to GUSTAVO PONCE, Department of Mathematics, South Hall, Room 6607, University of California, Santa Barbara, CA 93106; email: `ponce@math.ucsb.edu`

Partial differential equations and dynamical systems to PETER POLACIK, School of Mathematics, University of Minnesota, Minneapolis, MN 55455; email: `polacik@math.umn.edu`

Probability and statistics to RICHARD BASS, Department of Mathematics, University of Connecticut, Storrs, CT 06269-3009; email: `bass@math.uconn.edu`

Real analysis and partial differential equations to DANIEL TATARU, Department of Mathematics, University of California, Berkeley, Berkeley, CA 94720; email: `tataru@math.berkeley.edu`

All other communications to the editors should be addressed to the Managing Editor, ROBERT GURALNICK, Department of Mathematics, University of Southern California, Los Angeles, CA 90089-1113; email: `guralnic@math.usc.edu`.

Titles in This Series

946 **Jay Jorgenson and Serge Lang,** Heat Eisenstein series on $\mathrm{SL}_n(C)$, 2009
945 **Tobias H. Jäger,** The creation of strange non-chaotic attractors in non-smooth saddle-node bifurcations, 2009
944 **Yuri Kifer,** Large deviations and adiabatic transitions for dynamical systems and Markov processes in fully coupled averaging, 2009
943 **István Berkes and Michel Weber,** On the convergence of $\sum c_k f(n_k x)$, 2009
942 **Dirk Kussin,** Noncommutative curves of genus zero: Related to finite dimensional algebras, 2009
941 **Gelu Popescu,** Unitary invariants in multivariable operator theory, 2009
940 **Gérard Iooss and Pavel I. Plotnikov,** Small divisor problem in the theory of three-dimensional water gravity waves, 2009
939 **I. D. Suprunenko,** The minimal polynomials of unipotent elements in irreducible representations of the classical groups in odd characteristic, 2009
938 **Antonino Morassi and Edi Rosset,** Uniqueness and stability in determining a rigid inclusion in an elastic body, 2009
937 **Skip Garibaldi,** Cohomological invariants: Exceptional groups and spin groups, 2009
936 **André Martinez and Vania Sordoni,** Twisted pseudodifferential calculus and application to the quantum evolution of molecules, 2009
935 **Mihai Ciucu,** The scaling limit of the correlation of holes on the triangular lattice with periodic boundary conditions, 2009
934 **Arjen Doelman, Björn Sandstede, Arnd Scheel, and Guido Schneider,** The dynamics of modulated wave trains, 2009
933 **Luchezar Stoyanov,** Scattering resonances for several small convex bodies and the Lax-Phillips conjuecture, 2009
932 **Jun Kigami,** Volume doubling measures and heat kernel estimates of self-similar sets, 2009
931 **Robert C. Dalang and Marta Sanz-Solé,** Hölder-Sobolv regularity of the solution to the stochastic wave equation in dimension three, 2009
930 **Volkmar Liebscher,** Random sets and invariants for (type II) continuous tensor product systems of Hilbert spaces, 2009
929 **Richard F. Bass, Xia Chen, and Jay Rosen,** Moderate deviations for the range of planar random walks, 2009
928 **Ulrich Bunke,** Index theory, eta forms, and Deligne cohomology, 2009
927 **N. Chernov and D. Dolgopyat,** Brownian Brownian motion-I, 2009
926 **Riccardo Benedetti and Francesco Bonsante,** Canonical wick rotations in 3-dimensional gravity, 2009
925 **Sergey Zelik and Alexander Mielke,** Multi-pulse evolution and space-time chaos in dissipative systems, 2009
924 **Pierre-Emmanuel Caprace,** "Abstract" homomorphisms of split Kac-Moody groups, 2009
923 **Michael Jöllenbeck and Volkmar Welker,** Minimal resolutions via algebraic discrete Morse theory, 2009
922 **Ph. Barbe and W. P. McCormick,** Asymptotic expansions for infinite weighted convolutions of heavy tail distributions and applications, 2009
921 **Thomas Lehmkuhl,** Compactification of the Drinfeld modular surfaces, 2009
920 **Georgia Benkart, Thomas Gregory, and Alexander Premet,** The recognition theorem for graded Lie algebras in prime characteristic, 2009
919 **Roelof W. Bruggeman and Roberto J. Miatello,** Sum formula for SL_2 over a totally real number field, 2009